Introduction to BEEKEEPING

Introduction to BEEKEEPING

Jaydev Jha
Merin Joy
Shashank Vadlamani

2026
Agri Horti Press
New Delhi-110 084

ISBN: 978-93-49690-52-3 (Hardbound)

Published By

Agri Horti Press

New Delhi

E-mail: agrihortipress@gmail.com

Digitally Printed at : Replika Press Pvt. Ltd.

Preface

Beekeeping, an ancient practice that dates back thousands of years, has become increasingly relevant in today's world, not only as an agricultural pursuit but also as a critical factor in the global food chain. Bees, as essential pollinators, play a pivotal role in maintaining biodiversity, food production, and ecosystem health. As such, beekeeping is not merely a hobby or a profession—it is a vital contribution to the environment and agricultural sustainability.

This book, titled Introduction to Beekeeping, aims to provide a comprehensive guide to the art and science of managing honeybees. It serves as an essential resource for beginners and aspiring beekeepers, offering foundational knowledge on the biology of bees, hive management techniques, and the challenges and rewards of beekeeping. Whether you are interested in starting your own apiary, enhancing your existing beekeeping skills, or simply learning about the fascinating world of bees, this book offers valuable insights and practical advice.

The book begins by exploring the biology and behavior of honeybees, introducing readers to their complex social structures, communication systems, and vital role in pollination. A strong understanding of bee biology is critical for effective hive management and for ensuring the health and productivity of the colony. Following this, we delve into the practical aspects of beekeeping, from selecting the right equipment and setting up an apiary to understanding the different types of hives, managing swarming, and harvesting honey.

The importance of maintaining a healthy bee colony is a central theme of this book, and we examine common challenges faced by beekeepers, such as pests, diseases, and environmental factors. Special attention is given to sustainable and ethical beekeeping practices that support both the welfare of the bees and the ecosystem.

Furthermore, the book explores the myriad products derived from bees, such as honey, beeswax, propolis, and royal jelly, discussing their uses, benefits, and potential for value-added products. This holistic approach to beekeeping underscores its relevance not only for the environment but also for the economy, particularly in regions where agriculture and pollination services are closely intertwined.

Introduction to Beekeeping also touches on the global trends in beekeeping, including challenges related to declining bee populations and the critical importance of supporting bee conservation efforts. The book emphasizes the need for collaboration between beekeepers, scientists, and policymakers to ensure the continued health of pollinator populations worldwide.

Through this blend of practical advice, scientific insight, and ecological perspective, Introduction to Beekeeping aims to equip readers with the knowledge and skills needed to embark on a successful and rewarding beekeeping journey. It is our hope that this book will inspire a new generation of beekeepers to join in the essential work of protecting these extraordinary creatures while enjoying the many benefits that come with working closely with nature.

Jaydev Jha

Merin Joy

Shashank Vadlamani

Contents

Chapter 1

Understand Bee Biology and Behaviour

1.1 Understand different types of honey bees and their nature

1.1.1 Honeybees

- A honey bee is unsocial flying insect in the genus.
- These bees vary among themselves in traits like temperament, disease resistance and productivity.
- The environment has large effect on differences among bee colonies, but genetic makeup of colony also impacts characteristics that defines particular group.
- These are known for construction of perennial, colonial nests from wax, for large size of their colonies, and for surplus production and storage of honey.
- They distinguishes their hives as prized foraging target of many animals, like honey badgers, bears and human hunter-gatherers.
- The best known honey bee is western honey bee that is domesticated for honey production and crop pollination.
- They represents only small fraction of roughly 20,000 known species of bees.
- Many other types of related bees produce and store honey and have been kept by humans for the purpose of stingless honey bees, which is only member of genus Apis.

1.1.2 Origin

- Honey bees appear to have their center of origin in South and Southeast Asia as all extant species except Apis mellifera are native to that region.
- Living representatives of earliest lineages to diverge have their center of origin there.
- The first Apis bees appear in fossil record at Eocene-Oligocene boundary, in European deposits.
- The origin of prehistoric honey bees does not indicate Europe as place of origin of genus, only that bees were present in Europe by that time.
- As seen, no Apis species exist in New World during human times before A. mellifera by Europeans, but only single fossil species is documented ie, Apis nearctica.
- Honey bees constitute three clades ie Micrapis, Megapis and Apis.

1.1.3 Species and sub-species of honey bees

There are certain important species of honey bees such as:

- ❖ Rock bee, Apis dorsata (Apidae)
- ❖ Indian hive bee, Apis cerana indica (Apidae)
- ❖ Little bee, Apis florea (Apidae)
- ❖ European or Italian bee, Apis mellifera (Apidae)
- ❖ Dammer bee or stingless bee, Melipona irridipennis (Meliporidae)

1. Apis nigrocincta

- ❖ It is a honey bee species living on islands.
- ❖ Many colonies of such honey bee are found at elevations above 400 meters.
- ❖ The bees build medium-sized cavity nests with multiple combs that are nearly equidistant from each other.
- ❖ The nesting sites are close to the ground.
- ❖ Two types of combs are found with smaller being for workers and larger for drones.
- ❖ Cells for queen are found towards lower edges of combs.

2. Apis mellifera

- ❖ It is a European honey bee or western honey bee which is the most common among honey bee species.
- ❖ Such honey bee is highly social in nature.
- ❖ This species has been subdivided into at least 20 recognized subspecies (or races).
- ❖ It was one of the first insects to be domesticated and it the most common honey bee reared by beekeepers for their honey and role as a pollinator.
- ❖ This honey bee is currently found in all continents with exception of Antarctica.
- ❖ The honey bee is threatened by number of factors including pests and disease.
- ❖ It has haplo-diploid sex determination where unfertilized eggs develop in drones (males), and fertilized eggs develop into females.
- ❖ Female larvae that are fed standard diet of pollen, nectar and brood food become adult workers.
- ❖ Female larvae fed rich diet of royal jelly, pollen and nectar develop in queens.

- ❖ Worker honey bees are non-reproductive females which are smallest in physical size of three castes with their bodies specialized for pollen and nectar.
- ❖ Both hind legs of worker honey bee have corbicula specially designed to carry large quantities of pollen back to colony.

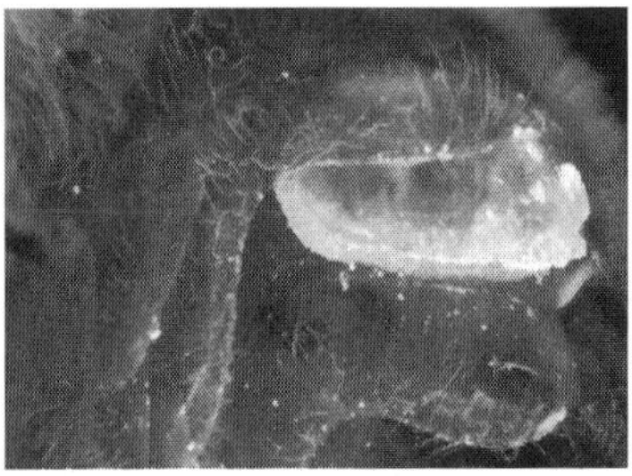

- ❖ Worker honey bees gives wax scales on underside of abdomen where scales construct the wax comb in the colony.
- ❖ Workers have barbed stinger that is torn with poison sac, from end of their abdomen when they deploy sting in tough-skinned victim.
- ❖ The queen honey bee is reproductive female in colony during normal circumstances.

- ❖ Drones are male caste of honey bees, where drone's head and thorax are more than that of female castes with large eyes that touches in top center of head.
- ❖ Their abdomen is thick and blunt at end which appears as bullet-shaped rather than pointed at end as with female castes.

3. Apis koschevnikovi

- ❖ It is a honey bee species that is found in Borneo island of Indonesia and Malaysia.
- ❖ These bees are all dark and brown banded.
- ❖ In this, queen and drones have light brown bands on abdomen while workers have light orange abdominal bands.

- These bees inhabit tropical evergreen forests.
- Habitat destruction due to clearance of land for plantations is main threat to survival of such bees.
- These bees live in small colonies in few combs in tree cavities in forests.
- It is a nectar-feeder and an important pollinator in its habitat.

4. Apis cerana

- It is known as eastern honey bee which is a honey bee species that lives in southeastern and southern Asia.
- The bees build nests with multiple combs in cavities with small entrance that protect colonies from threats.
- The nests are made up of 100,000 cells when complete and can house 6,000 to 7,000 individuals.
- Pollen, nectar, and honey constitute diet of such bees.
- It is a social species with eight distinct subspecies that are found in wide range of climatic zones, from taigas to moist tropical rainforests, steppes, grasslands, deciduous forests, etc.
- These honey bees are known for their ability to thermo regulate and use it as mechanism of defense.
- The physical characteristics of Apis cerana are similar to Apis mellifera where adult Apis cerana have branched hairs on their bodies to assist in pollen collection.
- The abdominal stripes of Apis cerana are pronounced to that of Apis mellifera as Apis cerana workers have four abdominal stripes, whereas Apis mellifera workers have three abdominal stripes.

- Worker body size varies among subpopulations all through Apis cerana's geographic distribution.
- The workers of southern subspecies of Apis cerana are smaller than Apis mellifera workers, but workers of northern subspecies of Apis cerana are larger than African subspecies of Apis mellifera.
- Queens are reproductive females of colony and are bigger than workers, with abdomens enlarged to fit their developed reproductive organs.

- Males of Apis cerana are called as drones.
- Each colony produces less drones than they do workers as drones have large eyes that meet at top of head, no stinger with thick and blunt abdomen at end.
- The thick and blunt abdomen gives drones more of blunt-ended body rather than pointed, aerodynamic body shape that is seen in female castes.
- Many colonies have single queen, thousands of workers, and many drones.
- The queen's only responsibility in colony to lay eggs and serves as mother of all workers in colony.
- The workers do all hive maintenance work related to tending to brood (eggs, larvae, and pupae), cleaning, foraging, and producing honey.

5. Apis dorsata

- These are very big honey bees that are found mainly in forest areas of South and Southeast Asia with significant populations in India, Nepal, Malaysia, and Singapore.
- Such bees are about 17-20 mm in length.
- The nests of such bees are built in open at great heights like cliff overhangs, tall buildings or tall trees in dense forests.

- These bees are known for their vicious behavior and aggressive nature.
- Their hives are important source of beeswax and honey which has been harvested for centuries by indigenous people of region inhabited by bees.
- The giant honey bees build single vertical comb that is covered by several layers of dense mass of bees.
- Male honey bees are drones which are identified as they have large eyes, short, round abdomens and lack stinger.
- Drones are produced during reproductive season which leaves their home colony to mate with queens from other colonies, where each colony produces more than thousand drones seasonally.

- Adult workers have short life span, as in other honey bee species, these workers have corbiculum on hind tibia and plumose body hairs that aid in pollen collection.
- The adult workers have a modified ovipositor that is used as a stinger.
- The conical shaped combs have hexagonal cells which are 1 meter wide.
- More than 200 colonies of 100,000 bees in each colony might build nests closer to each other, exhibiting high levels of social tolerance for each other.
- Humans have long relationship with giant honey bees, as paleolithic era cave paintings of such bees have been found in Valencia, Spain.
- The human induced activities sucha as pesticides, urbanization, honey hunting, and deforestation threaten bee populations in significant manner.
- These bees deliver painful stings and case of fatal defensive attack on humans exists.

6. Apis florea

- It is a red dwarf honey bee which is small type of honey bee species that is found in parts of southern and southeastern Asia and Africa.
- The bees build open nests and form small colonies as it makes susceptible to predators.

- These bees dance to reach decision to build new nesting site.
- The decision is taken when majority of individuals dance in direction of proposed new site.
- The worker bees uses auditory signal of piping to inform swarm whose decision has been taken and entire swarm moves off in new direction.
- Such bees build single comb nest on single branch and salvage wax from previous nest if close to new one.
- These bees feed on honey, nectar and pollen.
- Such bees communicate foraging signals by dancing displays.
- The red dwarf honey bees are hunted for their honey.
- The bees have important place in Hindu and Buddhist culture.
- Due to increased deforestation, forces the bees to restrict their range and occupy areas that are less populated.
- Younger Apis Florea works in the nests and do maintenance while older ones are used for protection and foraging.
- They expand to new place by human introduction or itself and poses threat to native insect like Apis. mellifera as they compete with same resources.
- Nests of A. florea are exposed and made up of single horizontal comb that is built around and attached to tree branches or other support.

7. Apis and reniformis

- It is a black dwarf honey bee which lives in tropical and subtropical areas of southeast Asia.
- It is rare species which has been classified as separate species.
- Such bees are darkest of Apis genus, with its queens are entirely dark.
- The worker bees have light and yellow scutellums.
- These bees form open-air nests in darker areas of forest which receives 25%-30% of normal sunlight.
- The nests are single-combed and hang from small twigs of forest trees at altitude of around 2.5 meters.
- The honeycombs are about 70-90 mm in size.
- Honey storage is found in area surrounding the branch and below it lies broad area where honey bee larvae develop.

- The black dwarf honey bee is lowland species which migrates to higher lands during rainy season.
- The species is used by humans for commercial products like honey, wax, royal jelly, and bee venoms.
- Such bees are important pollinating agents in their region of occurrence.

8. Dammer Bee

- Apart from true honey bees, two species of stingless or dammer bees like Melipona and Trigona are present in the country in abundance.
- These bees are smaller than true honey bees and build irregular combs of wax and resinous substances in crevices and hollow tree trunks.
- The stingless bees have importance in pollination of food crops.
- They bite their enemies or intruders.
- They can be domesticated, but honey yield per hive per year is only 100 Gms.

1.2 Understand life cycle of different types of honey bees

1.2.1 Life span of different bees

- Bees play important role in ecosystem and honey bee is most important single species of pollinator, which ensures people and animals to have food they need to survive.
- If there were no honey bees, world wouldn't have broccoli, berries, apples, cucumbers and other foods.
- Bees also produce honey and wax while honey bee has short life span.
- Bee colony which is organized, sophisticated society made of three categories, ie single fertile queen bee, male drone bees and sterile female worker bees.
- Bee's caste and time of year in which they born, affects the lifespan.
- Summer workers have shortest honey bee lifespan, while queen bee outlives both other castes.

Life Span of Drone Bees

- Adult drones have no useful purpose in bee hive.
- They do not provide food, feed the young or produce wax.
- They waste colony's resources and serve to mate with queen bee.
- They were the first to leave the hive in six days after emerging from pupal cell and flies to areas and goes back to hive only when they failed to mate.
- Successful maters die minutes or hours after mating with queen while rest of drone bees survive only as long as worker bees allow them to.
- If there is a shortage of food, worker bees kill or kicks out drones.
- These bees rarely survive the winter, as worker bees want to protect their limited resources.

- When drone bee is ejected from hive, they soon dies from cold or starvation.
- The average life span of drone bee is 8 weeks.

Life Span of Worker Bees

- Initial part of worker's life is spent working in hive, while last part is spent in finding the food and gathering pollen or nectar.
- Worker bees gather water to cool the inside of nest on hot days, and use water to dilute honey before feeding it to larvae.
- It is worker bees who are responsible for pollination and when they land on plants or flowers, they collect pollen dust all over their bodies, and use their adapted legs to discard the pollen, leaving it on other plants.
- During summer, worker bees live for 5-6 weeks, due to their heavy workload which makes better of them as its is most active time of year, when they spend their days foraging for food, storing nectar, feeding larvae and producing honey.
- These bees live longer in winter as their fat supply increase and their glands provide food for larvae.

Life Span of Queen Bees

- Queen bee has very important function in the colony, and has longest life span.
- As average life span of queen bee is 2-5 years, these bees have known to live up to 7 years, as this is rare.
- About a week after new queen emerges from cell, queen bee goes on many flights for mating with as many as 20 drones.
- After queen bee returns to lay her eggs, they rarely leave the colony and lays between 1,000 and 2,000 eggs per day in the hive.
- If queen bee fertilizes egg, egg will become female as worker or queen bee, while if queen bee does not fertilize egg, it becomes male drone bee.
- The queen's survival in winter months depends on viability of her colony.
- A strong group of worker bees protects the queen and regulates her temperature.
- The worker bees keep close eye on queen bee to make sure that queen bee is up to her job.
- If queen bee doesn't lay enough eggs, workers will start developing new queen to replace the old one through supersedure process.
- The new queen is pampered with food and affection, while old queen is neglected and left to waste away.
- In some beekeeping practices, the beekeeper replaces the queen after one or two years.

Factors Affecting Honey Bee Life Span

- The life span of bee can be affected by many factors such as:
- Bees die from natural causes, but they can be eaten by other animals or killed by other bees.

- Worker bees die due to overwork, but biggest threat to bees is disease or infection, that destroys entire colonies.
- The parasitic fly Apocephalus borealis forces bees to leave hive and die, and then fly where larvae emerges from dead bees.
- The flying of larva spreads deformed wing virus.
- Other threats to honey bees are pesticides, habitat loss and mites.

1.2.2 Life cycle

- Honey bees pass through distinct life stages such as egg, larva, pupa and adult.
- The process is called complete metamorphosis as the form of bee changes drastically from larva to adult.
- Passing through immature stages takes 21 days for worker bees.
- On first day, queen bee lays a single egg in each cell of comb where egg hatches in larva on fourth day.
- The larva is legless grub that resembles a tiny white sausage.
- The larva is fed a mixture of pollen and nectar called as beebread.
- The larval stage is similar for worker bees and queen bees, which are both female bees that hatch from fertilized egg.
- Worker bees, drone bees and queen bees are all fed royal jelly in their first few days as larvae, but after that only the queen bee continues to receive the royal jelly, supplemented with honey toward the end of the larval stage.
- Worker bee larvae are mass-fed a compound known as "worker jelly" or "brood food," while drone bees, male bees that hatch from unfertilized eggs, are fed a modified version of the worker bees' diet, including increased quantities of pollen and honey, during the larval stage.
- On ninth day, cell is capped with wax and larva is transforms into pupa.
- The pupa which doesnt eat is a physical transition stage between amorphous larva and hairy winged adult.
- On 21 day, new adult worker bee emerges.
- During pupal stage, bees form wings, legs, internal organs and other adult body parts, using stores of accumulated fat during larval stage.
- Small hairs grow on bee's body and total time it takes to become fully developed adult bee is 21 days for workers, 24 days for drones and 16 days for queen bees.

Features:

- Eggs are laid singly in a cell in wax honeycomb that produces and shaped by worker bees.
- Using her spermatheca, queen bee choose to fertilize egg she laid depending on type of cell she is laying.
- Drones develop from unfertilised eggs which are haploid, while females develop from fertilised eggs and are diploid.

- Larvae are fed with royal jelly produced by worker bees that switches to honey and pollen.
- The exception is larva which fed mainly on royal jelly that develops into queen bee.
- The larva undergoes many moultings before spinning cocoon in the cell, and pupating.
- Young worker bees, called as nurse bees, clean the hive and feed the larvae.
- When royal jelly-produces glands it starts to atrophy and begin building comb cells which progresses to other in the colony work as they become older like receiving nectar and pollen from foragers and guarding hive.
- The worker takes her first orientation flights and finally leaves the hive and typically spends the remainder of her life as a forager.
- Worker bees cooperate to find food and use pattern of dancing to communicate information regarding resources with each other, that varies from species to species, while all living species of Apis exhibit certain form of behavior.
- If resources are very close to hive, they exhibit less specific dance known as round dance.
- Honey bees do tremble dances, which recruit receiver bees to collect nectar from returning foragers.
- Virgin queens go on mating flights away from home colony to drone congregation area, and mate with many drones before returning.
- The drones die during act of mating, so queen honey bees do not mate with drones from their home colony.
- Colonies are established not by solitary queens, but by groups known as swarms, which has mated queen and large contingent of worker bees.
- Such group moves en masse to nest site which was scouted by worker bees beforehand and the location gets communicated with special type of dance.
- Once swarm arrives, it immediately construct new wax comb and begins to raise new worker brood, so the type of nest founding is not seen in any other living bee genus, as many groups of vespid wasps found new nests by swarming.
- Stingless bees start new nests with large numbers of worker bees, but nest is constructed before queen is escorted to site, and such worker force is not true swarm.

Life Cycle of Apis mellifera

- With highly social life history, Apis mellifera honey bee colonies are considered as super organisms where entire colony is viewed as biological unit.
- The honey bees reproduce not by producing more individual bees, but rather by producing more colonies, where reproductive process of creating new colony is called as swarming.
- European honey bees swarm in spring and early summer when pollen and nectar resources are plentiful, so to start swarming process, 10-20 daughter queens are produced by colony.

- If daughter queens are in late pupal stage, mother queen and about 2/3rd of adult workers leaves colony and travel to location where they coalesce as while they send scout workers in search of place to establish new colony.
- The daughter queens in original colony emerge as adults and fight till single queen remains alive, unless one queen emerges earlier than her sisters, where she hunts for and kill unborn siblings.

- After short time of maturation, remaining daughter queen, leaves the colony to mate with 15 drones.
- All mating occurs in first two weeks of queen's life, outside the hive in an air at drone congregation areas where queen stores collected sperm in her spermatheca for remainder of her life.
- After mating is done, daughter queen begins laying eggs and completes halves of swarming process.
- European honey bees are adapted to temperate climates, where there is short season with generous amounts of pollen and nectar available, due to which, they swarm only once a year.
- The remainder of spring/summer is devoted to collecting and storing required nectar and pollen to generate food stores for surviving fall and winter, where behavior of resource hording makes European honey bees good honey producers.
- Nectar is collected from flowers and transformes in honey by enzymatic processes and dehydration within the colony, where honey is capped in wax comb from where, it stays fresh indefinitely, as per original nectar source.
- Beekeepers and honey hunters collect honey comb for human use.

Life Cycle of Apis dorsta

- All honey bee species are superorganisms in whole colonies, rather than individual bees, are considered as biological unit.
- In this, one colony can split into two colonies once the colony becomes large in size and there are abundant food resources.

- Such colony-level reproduction is known as swarming.
- The lifespan of Apis dorsata workers is not well characterized.
- Compared to Apis mellifera workers, Apis dorsata seems to live significantly longer, especially during migration swarms when workers will sometimes travel more than two months to reach a new destination and produce a new generation of bees.

The life cycle and life stages of Apis dorsata are same as those in all other honey bee species as honey bees are holometabolous insects which undergo four separate life stages:

- egg
- larva
- pupa
- adult

Egg:

- A queen will lay one egg in an individual wax cell at a time.
- Eggs are small, white, and oval; they look similar to a grain of rice.
- After three days, a larva will hatch from the egg.

Larva:

- The larvae are small, white, and maggot-like.
- It remain in their individual cells.
- Larva are fed a steady diet of a protein-based liquid produced by younger adult workers.
- For six days, the larvae will eat until their body weight attains that of a fully developed adult.
- They are usually found lying in the bottom of a wax cell in the shape of a letter C.

Pupa:

- Once the larvae are ready to pupate, the workers cover the cell with a wax capping.
- The pupa will remain undisturbed for 12 days.

Adult:

- After completing the final molt and enclosing from the pupal casing, the adult bee will exit the cell by chewing through the wax capping.

Life Cycle of Apis florea

- Eggs are laid singly in cell in a wax honeycomb which are produced and shaped by worker bees, as queen fertilizes egg she lays.
- Drones develop from unfertilized eggs while females develop from fertilized eggs.
- Larvae are fed with royal jelly produced by worker bees which later switches to honey and pollen.
- The larva since inception is fed solely on royal jelly, which develops into queen bee. Further, larva undergoes many moltings before spinning cocoon in cell and pupating.
- Young worker bees clean the hive and feed the larvae. When their royal jelly-produces glands begins atrophy by begining building comb cells which progress to other in colony tasks as they become older like receiving nectar and pollen from foragers, and guarding hive.
- In this, worker takes her first orientation flights and finally leaves the hive and typically spends the remainder of her life as a forager.
- Development from egg emerges bee which varies among queens, workers and drones. Queens emerge from cells in 15-16 days, workers in 21 days and drones in 24 days.
- The honeycomb belongs to species of honey bees called as Dwarf Honey bees where stings are incapable of penetrating human skin.

Life Cycle of Apis cerana indica

Apis cerana is holometabolous which undergoes four distinct life stages:

- egg,
- larva,
- pupa,
- adult

Eggs:

- The queen lays a single egg in each brood cell.
- Eggs are small, white, and oval in shape.
- Larvae emerge from eggs after three days.

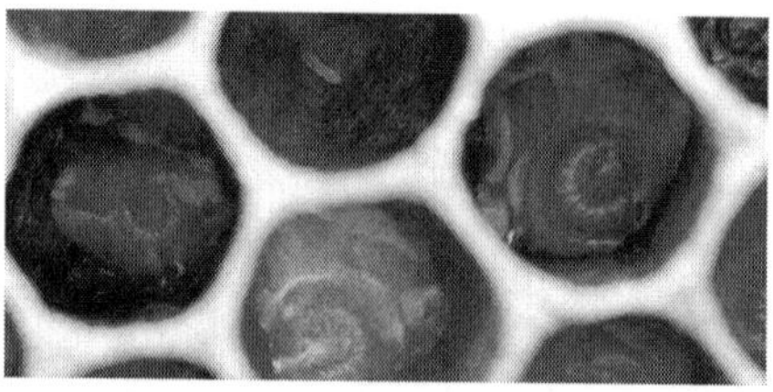

Larvae:

- Newly hatched larvae curl into shape of C at bottom of cell.
- Larvae are white in color, blind, and legless, with wet shine.
- The larvae are fed brood food and/or royal jelly in their cells till they are large enough to pupate, at which time the adult worker bees cap the larval cells.

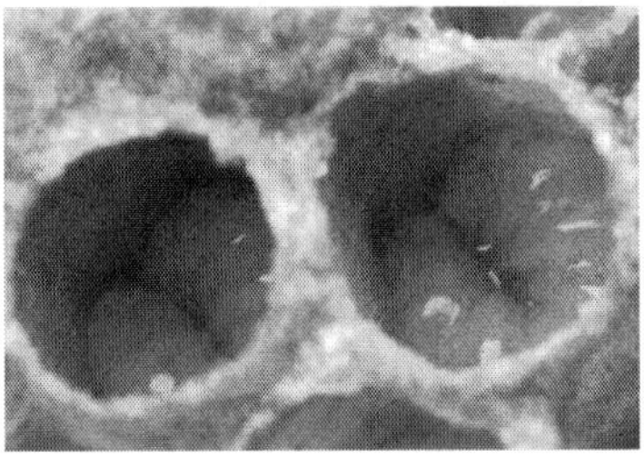

Pupae:

- The larvae molt into pupae beneath the wax cappings and pupal cells are undisturbed until pupation is complete.
- The wax capping of drone pupal cells in Apis cerana have distinctive pore.
- The honey bee, Apis koschevnikovi Buttel-Reepen, constructs similar pore on drone pupal capping.

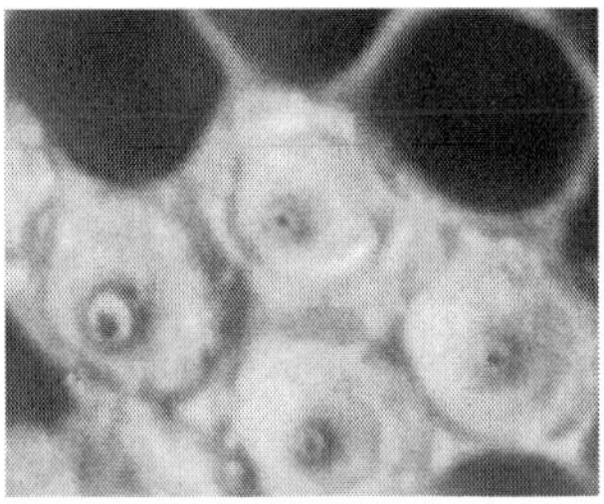

Adults:

- Following their pupal development, new adults chew their way out of their capped cells.
- Honey bees are considered superorganisms, wherein whole colony is biological unit rather than individual bees.
- Colony-level reproduction is swarming and happens in spring and summer, but swarming may be more frequent in tropical areas where climate is more favorable year-round.
- Abundant resources and large colony size are main triggers for swarming.
- To initiate swarming, colony will raise 10-20 daughter queens and when daughter queens are in late pupal stage, mother queen and up to 2/3rd of adult workers leave colony in search of a place to have new colony.

Life cycle of Apis mellifera

Introduction

- Apis mellifera are red/brown with black bands and orange yellow rings on abdomen having hair on thorax and less hair on abdomen.
- They have pollen basket on their hind legs.
- Honeybee legs are mostly dark brown/black.

- There are two castes of females, sterile workers are smaller, fertile queens are larger.
- Males called as drones, are 15-17 mm long at maturity, as smaller workers have longer wings than drones.
- Both castes of females have stinger that is formed from modified ovipositor structures.
- In workers, sting is barbed, and tears away from body when used.
- In both caste, stinger is supplied with venom from glands in abdomen.
- Males have much larger eyes than females, probably to help locate flying queens during mating flights.
- There are 26 recognized subspecies of Apis mellifera, with differences based on differences in morphology and molecular characteristics.
- The differences among subspecies is discussed in terms of agricultural output in particular environmental conditions.
- Many subspecies have ability to tolerate warmer or colder climates.
- Subspecies vary in their defensive behavior, tongue length, wingspan, and coloration. Abdominal banding patterns also differ - some darker and some with more of a mix between darker and lighter banding patterns.

Development

- Honeybees build a hive out of wax secretions from their bodies, and queens lay their eggs in cells in the wax.
- The speed of subsequent development of young is affected by temperature and is fastest at 33-36°C.
- Honeybees are holometabolous insects having four stages in life cycle such as egg, larva, pupa, and adult.
- mellifera eggs hatch in 28-144 hours, as per their temperature.
- The larva which emerges is small white grub which stays in wax cell that grows which is fed and groomed by adult workers.
- The food that female larva receives shows whether it will be a queen or worker.
- At 34°C, larvae feed and grow for 4-5 days, queens for 6 days, and males for 6-7 days.
- At end of that period, their cell is sealed by adult workers, and larva molts, spins silk cocoon, and transforms into pupa stage.
- Pupae undergo big metamorphosis that takes about 7-8 days for queens, 12 days for workers, and 14-15 days for males.
- Once final metamorphosis is complete, they chew their way out of cell and begin their adult life and will not grow or molt after emerging.
- Adult workers will live for 2-4 weeks in the summer, or as long as 11 months if they live through winter.
- Males only survive for 4-8 weeks, and do not live all through the winter.
- Queens live 2-5 years.

- In larval stage, larva is fed with royal jelly, pollen/nectar, and honey combination and goes in pupae stage where it caps itself in cell to metamorphose into mature stage.

Reproduction

- The great majority of female A. mellifera in hive are sterile workers where only queens mate and lay eggs as there is only single reproductive queen in hive.
- During periods of suitable mild weather in spring and summer, males leave hive and gather at drone assembly areas near hive.
- Virgin queens fly through areas, attracting males with pheromones where males pursue and attempt to mate with queen in flight.
- The comet forms, as cluster of males forms around female with string of other males that tries to catch up.
- Every male who succeeds in mating drops away and dies in few hours or days.
- Males who do not mate will continue to loiter in assembly areas till they mate or die trying.
- Queens will mate with up to 10 males in a single flight and will mate with males from their own hive, or from other hives in area.
- The queen's mating behavior is centered around finding best place to mate beforehand, by taking directional flights for period of time, lasting no more than couple of days.
- Afterward, queen leaves the hive and flies to mate with drones in assembly area which starts to occur after first week of birth.
- The queen does mating up to four times and after this congregate of mating, queen never mates again in lifetime.
- Apis mellifera queens are main reproducers of nest and all activities of colony are centered around their reproductive behaviors and survival.
- The queen is only fertile female in colony which lays eggs nearly continuously throughout the year, sometimes pausing in late fall in cold climates.
- A fertile queen lays as many as 1,000 eggs/day, and 200,000 eggs in her lifetime which takes a queen about 16 days to reach adulthood, while in another week or more to begin laying eggs.
- Males take about 24 days to emerge as adults, and starts leaving the nest for assembly areas a few days after that.
- Queen honeybees control whether or not an egg lay is fertilized as unfertilized eggs develop as males are haploid while fertilized eggs are diploid and develop as workers or new queens, as per how they are fed as larvae.
- Queens increase the ratio of male to female eggs they lay if they are diseased or injured, or in response to problems in colony.

1.3 Study bee behavior (communication, swarming, absconding, robbing etc.)

1.3.1 Communication

The honey bees live in a colony and communicate with each other by passing information using various pheromones. The worker bees communicate information by using their peculiar dance which can be in shape of:

(a) Round dance:

- ❖ It shows short distance of food source.
- ❖ In this, bees run in circles, initially in one direction and then in opposite direction.
- ❖ If source of food is less than 100m away, then bees perform round dance making small circles.
- ❖ The number of circles formed by bees shows the distance of beehive from food source.
- ❖ The workers dance vigorously, if quantity of food is more and of superior quality.
- ❖ If food is inferior and quality is less, then dance is slow and shorter.

(b) Tail wagging dance or Wag-tail dance:

- ❖ It is used to show long distance of food source where bee makes two half circles in opposite directions with straight run in between.
- ❖ During straight run, bee shakes its abdomen from side to side, where number of wags per unit time is inversely proportional to distance of food.

1.3.2 Swarming

Introduction:

- ❖ It is a behavior that takes place among bees in April to June.
- ❖ It was considered good as beekeepers were able to increase the number of colonies by capturing swarms.
- ❖ It is now considered as nuisance since it instantly reduces honey production.

Causes

- ❖ The causes of swarming results due to:
- ❖ over-population or congestion in hive.
- ❖ imbalance between old worker bees and young worker bees.
- ❖ over-heating of hive, as a result, bees are unable to adequately ventilate the hive by placing langstroth hives in places where there is absolutely no shade around.
- ❖ no longer serving of hive structure at full capacity as a result of over-abundance of defective cells.
- ❖ queen bee unable to lay eggs as many new cells have already been filled with cured nectar or pollen.

- inclement and inhospitable weather, which does not allow bees to exit the hive.
- queen which unable to lay eggs because of sickness or in state of dying.
- non production of required pheromones by queen bee to keep the colony together.
- genetics and bee's race.
- presence of idle nursing bees which signals internal problem that leads to swarming.

Preparation of Swarm

Regular inspection is required to spot possible swarming preparation or swarming behavior, so it is recommended to bee keeper to inspect bee hives at least once per week. The following are certain clues that colony may be preparing to leave bee hive which includes:

- Visible increase in population of female worker bees in colony.
- Drones are suddenly reared in large numbers.
- Hive expansion is no longer taking place even if cells are full to capacity with pollen, larvae and honey.
- Queen cells or queen cups are seen near bottom sides of frame.
- Queen bee already laid eggs in queen cups.
- Normal egg production of queen bee slow down, as number of brood being reared by nursing bees decreases.
- Queen bee appears restless and is not doing what she usually does.
- The open queen cells has multiple larvae where larvae may or may not differ in age.
- Field bees or foraging bees are no longer active as they used to be, but it is possible to see some of these field bees hovering near hive.
- Swarm cells or queen cups are sealed by workers.
- With all above, preparation ends and swarm is finally cast.

Preventive Measures

Every beekeeper wants to keep the hives intact. Swarming is no longer considered an efficient, natural way to reproduce colonies. There are certain steps which bee keeper needs to be considered to prevent bees from swarming which includes:

- Congestion is the main reason that shows why bees decide to swarm, so beekeeper can easily ease the congestion in a colony by adding more room to the hive.
- By adding brooding and storage frames, queen bee will able to resume bee egg-laying duties where workers will resume collecting nectar and pollen.
- Temporarily separate the queen from rest of colony till all issues in hive has been resolved, which can be done by keeping queen in special super/frame at topmost section of hive, to stop the queen from flying out.
- Reverse the hive to ease congestion due to winter clustering.

Preventing Swarming Behavior

- Colony splitting involves in division of colony population in half where old queen is taken to new hive with half of colony and population in old hive that allows to raise new queen.
- The old queen and half of population needs to be installed at distant hive and such technique works and makes honey production to reduce in process.
- Supering can be installed to hive so that bees have extra space to store pollen and honey, so bee keeper needs to replace queen bee every two years where colonies with young queens are less prone to swarm.
- Colonies with queen bees are 3 years older and are twice or thrice to exhibit swarming behavior which colonies with young queen bees.
- Beekeeper should avoid supersedure completely which happens when queen bee is no longer laying as many eggs as before or when queen bee is not producing enough pheromone, so colony often produces supersedure cells or additional queen cells.
- When new virgin queen bee emerges, fight will ensue and when old queen bee gets killed or driven out by new queen bee, risk for swarming gets doubled.

1.3.3 Robbing

Introduction

- Honey bee workers invade and steal honey/nectar from other colonies or sugar/corn syrup from feeders and deliver syrup to other colonies, which is called as robbing behavior.
- Robbing behavior involves collection of nectar and honey, but not pollen or brood.
- Many beekeepers shows that robbing bees steal wax or propolis from other hives.
- The behavior can escalate quickly from just few bees robbing other colonies to massive frenzy of bees by robbing many colonies in apiary.
- In normal foraging behavior, honey bees gather nectar from blooming flowers in their foraging range and pass the nectar to bees who deposit it into wax cells in the hive.
- The nectar is processed into honey, capped over with wax, and stored in the hive.
- Many nectar are combined with pollen and resulting product stored as bee bread, a food for the brood and young adult bees in hive.
- Nectar flow or honey flow is a period when ample local plants are blooming and producing nectar that can be collected by bees.
- Honey bees are less likely to rob other colonies when nectar is available in environment while bees are likely to rob other colonies when nectar resources are scarce.
- Weak colonies are prone to being robbed as they are unable to defend themselves from the onslaught of invading bees.
- Robbing occurs between hives in one apiary or between hives in different apiaries and which happens between multiple colonies simultaneously.

- It is common during robbing frenzy to see colonies robbing other colonies and being robbed at the same time.
- The bees involve in robbing are older bees that otherwise would be foragers, but non-foraging bees can become robbers when robbing behavior reaches frenzy level.
- Robbing increase in intensity in short amount of time as robber bees return to their nest and recruit more robbers from their own hive to join in the behavior.
- Robbing serves as main problem during summer or when other major nectar deaths occur.
- Any apiary can experience robbing at any time of year, although certain situations promote robbing more than do others.
- Robbing promotes significant bee drift, or movement of individual bees from their home colonies to other colonies which increases risk of transferring pests and diseases between colonies.
- During robbing events, there are high number of bees flying around the apiary and actively scouting around the outside of nearby colonies.
- There are bees fighting at colony entrance and/or trying to gain entry in colony by areas where two supers meet and/or around the hive lid and/or bottom board.
- Colonies in apiary have heightened defensive response, and there are more dead bees at colony entrance.
- There masy be remnants of shredded wax cappings or other hive debris near or under the colony or colonies being robbed where debris may be on hive bottom board.
- Bees will continue to rob a source till it is depleted, thus a colony can be robbed of all of its honey/nectar resources in a matter of hours or days.

Occurrence

- Robbing occurs throughout daylight hours, usually during a nectar dearth.
- In this, colony bee populations grow during nectar flows.
- The cessation of nectar flows leads to colonies with high populations, many forager bees, and few available nectar resources in the environment, so old forager bees become attracted to weaker hives from which they can steal nectar and honey resources.
- Weak colonies have proportionately fewer guard bees protecting their entrance and resultantly are more prone to be robbed.
- It is important to recognize periods of nectar dearth in order to anticipate the initiation of robbing behavior.
- Robbing bees invade other colonies by nest entrance or through any joints, holes, cracks, or other openings in hive.
- Beekeepers working their colonies during nectar dearths make the colonies vulnerable to robbing bees.
- Large apiaries with colonies located close to one another are more prone to robbing than are smaller apiaries with fewer, well-spaced colonies.

- Robbing bees will exploit any sugary resource in proximity to their nest which includes honey buckets, jars or jugs of sugar syrup, supers left unattended, hives left open or simply not closed properly, honey extraction equipment, and even honey processing rooms or facilities.

Signs of robbing

- A colony that is under attack will appear active than it was previously, but activity at entrance is different.
- Bees when they are coming back to their own home land on the periphery of the entrance then walk in.
- Bees that are robbing will dart around at the entrance, then make a dash for it, but straight for the entrance, not landing on the periphery as bees from the colony will.
- Defending bees will become agitated and there will probably be fighting until they are overwhelmed, when they will simply give up.
- If beekeeper take the crownboard off, robber bees will fly out in a hurry.
- Capped honey will have the cappings ripped off and the holes will have jagged edges.
- The robbing colonies will be more active than usual, but there won't be the signs that you get in a robbed colony.

Preventing Robbing

- Maintain queenright colonies of uniform strength and in good equipment.
- Work colonies carefully but quickly during nectar dearths.
- Use an entrance reducer on all colonies during times of nectar dearth.
- Always keep supers covered when removing them to work a colony.
- Do not work more than one colony at a time.
- Remove any excess supers or honey that the bees in a colony are unable to protect.
- Equalize colonies (i.e. take steps to make them similar in strength).
- Move weaker colonies to another location.
- Avoid dripping syrup, nectar, or honey on the outside surfaces of or the ground around hives.
- Work colonies later in the day to reduce the amount of daylight hours bees have to rob.
- When feeding colonies, always keep external feeder surfaces free of syrup and clean up any syrup spills.
- Do not leave wax, honey debris, or combs exposed in the apiary.

1.3.4 Absconding

- Removal of any colony without emergence of new queens or over-crowding of bees is considered as absconding.
- In this, all colonies are observed weekly and number of colonies absconded in nesting sites during study period was recorded.

- When colony swarms, it splits in two parts, where one part stays in old home and other part finds a new home.
- Absconding bees leaves the hive completely to live somewhere else and can leave the hive anytime that bees are active, even in fall.
- In this, colony may be under attack from predators at night where skunks and opossums eat honey bees and scratch at hive entrance during night.
- The tendency to abscond is determined by climate and effects of climatic change on flowering and nectar flow.
- Honey bees have possible strategies for surviving periods of dearth by storing sufficient quantities of honey to allow them to survive during dearth period or by moving to other place where there may be more nectar flow.
- The absconding can be planned and unplanned.
- Absconding rates are on average 15-30% annually and can be high as 100% in some seasons.

1.4 Study working behavior of different castes of honey bee

1.4.1 Introduction

- Honey bees are social creatures which enlist caste system to do the work that ensures survival of colony.
- Many worker bees all sterile females, are responsible for feeding, cleaning, nursing and defending the group.
- Male drones live to mate with queen, who is fertile female in colony.

1.4.2 Casts of Honey bees

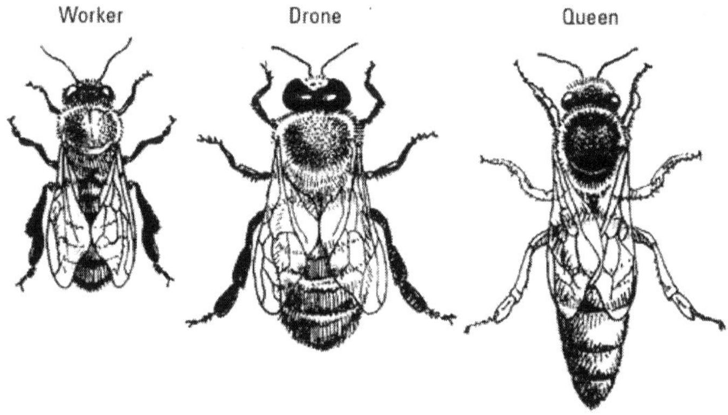

- There are three types of individuals in a colony, namely the Queen, worker and drone.
- All these three castes depend on each other for their existence.
- Honey bee colony has single queen, few hundred drones and many worker castes of honey bees where queen is fertile, functional female while worker is sterile female and drone is male insect.

- Due to existence of many morphological forms, bees are polymorphic species.

1. Queen:

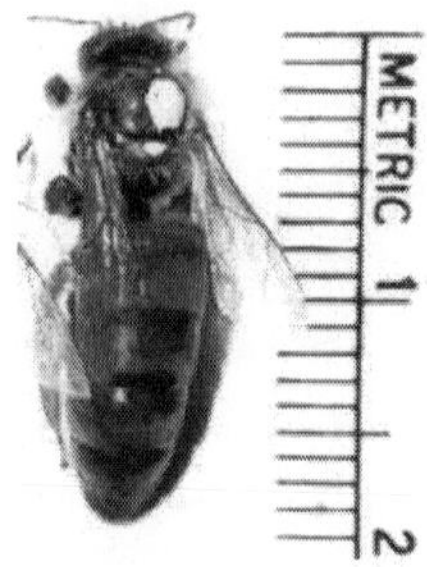

- It is diploid, fertile female where there is single queen in a colony that is larger than members of other castes.
- The presence of queen in a colony is required.
- The size of queen body is larger than other castes of bees of colony.
- The legs are strong as she always walks about on the comb.
- It has sting curved like scimitar at tip of the abdomen, which is a modification of egg-laying organ known as ovipositor.
- The sting serves as organ of defence.
- Queen never uses it against anybody except her own caste.
- It is responsible for laying eggs for a colony which lays 1000-1500 eggs everyday and lives a life of two three years.
- The number of eggs laid per day varies from individual to individual, but production is as much as 6,000 eggs per day.
- The queen lays fertilized eggs from which females develop and unfertilized eggs from which males develop.

2. Worker:

- It is a diploid which is a sterile female.
- The size of worker is smallest among castes which constitute the majority of bees in colony.

- The function of worker is to collect honey, to look after young ones to clean the comb, to defend the hive and to maintain the temperature of the hive.
- Many adaptations occurred in worker for doing her various functions.
- Its body is covered with branched hairs so that if bee visits flower, pollen grains adhere to hairs and other parts of body.
- The worker cleans off pollen grains with special structures, antenna cleaners on each foreleg, pollen brushes on all legs and pollen combs on hind legs.
- All pollen is kept in pollen basket present on the outer surface of tibiae on hind legs.
- Water and nectar are gathered by using sucking mouthparts which are modifications of maxillae and labium.
- Workers are provided by sting at tip of abdomen that shows modified ovipositor.
- In this, large poison storage sac is connected with base of sting.
- In this, two acidic and one alkaline gland mix their secretion to form poison which is injected by operation of muscles to other animals.
- During withdrawal from prey's body, the stings along with other poison apparatus are torn off, resulting into death of particular bee.
- Workers are female but are incapable of producing eggs.
- The life span of worker bee is between 4-5 months but during hard working days, they persist for 5-6 weeks only.

3. Drone:

- It is a male bee that is the product of an unfertilized egg.
- It has bigger eyes and lack stingers.
- They are larger than workers and are quite noisy.
- They are unable to gather food, but eat voraciously.
- Their main function is to fertilize the queen bee.
- They are incapable for doing ordinary work of hive.
- They help in maintaining hive temperature.
- They go out of hive at mid-day when weather is warm.
- The number of drones in a colony varies from 200-300, but during bad season they are driven out.
- The drone develops parthenogenetically from unfertilized eggs.

1.5 Apis mellifera in hive and in field

- Apis mellifera adansonii is small having its weight as 85 mg.
- Its cell diameters is 4.7 mm where bees colouration is not uniform.
- Many colonies are hybrids of yellow and black where black variety is generally less aggressive but absconds readily and has better capacity for regulating inner hive temperature when weather is cold.
- These are fast when leaving the hive and begins work half an hour before sunrise to collect the nectar produced during cool hours of night.
- In this, hive moved 5-10 m during the night and after foraging it, will come back to the original hive place circle few times and join the hive in its new position.
- It is an important drift that shows why local beekeepers always place hives at about 3-4 m apart.
- A hive is provided which can be opened and examined without unduly disturbing the colony.
- The hive is fitted with vertical wooden frames in which the bees can build their combs.
- The hive frames have wired in their centre, a sheet of wax which is indented with a hexagonal pattern so that the workers build their combs in confined of frame, where each comb can be removed separately.

Chapter 2

Handle Bee-Keeping System and Bee-Keeping Tools

2.1 Understand the difference between traditional and modern beekeeping system

2.1.1 Introduction

Honeybees nest in open produces less honey as compared to those, who nest in confined enclosures as colonies in open are exposed to predators, which uses many workers as guards to fight against intruders. They have to consume large quantities of honey, which they use as fuel, to allow them to cluster to stop wind which cools down their combs and to generate enough heat to maintain proper temperature for brood development.

During hot days, more bee uses honey as fuel to allow to fan to cool melting combs in order to avoid disaster. Such temperature controls can be quite inefficient, as colonies are exposed to condition. The exposed colony has to keep larger numbers of house bees, where less foraging bees are present to bring in required nectar and other essentials from the field.

The types of hives and way of keeping bees varies from area to area, so based on locally available materials used for construction of hives, environmental conditions and positions used to keep bees, the following variants of basic design are found throughout the country such as hollowed logs, bark hive, bamboo or reed grass hive, mud/clay hive, animal dung hive, woven straw hive, gourd hive, earthen pot hive etc.

2.1.2 Traditional hives systems

- ❖ Beekeeping has been practiced from time immemorial where large areas having wooden boards or timber are scarce, so grass and mud played major roles in providing material for beehive construction.
- ❖ Traditional beekeeping uses cheap and plentiful local materials for hive construction, some of which would otherwise be wasted, e.g. ciba palm, but such simple beehives cannot be easily manipulated as bees fix combs to hive body.
- ❖ Combs cannot be inspected at all and detached combs cannot be easily replaced, so to counter such problems, traditional beekeepers should adopt top-bar system as simple modern way to convert traditional hives into movable-comb hives.

- -Beekeeping started with traditional or fixed comb hives as combs are attached to top and sides of hive itself and beekeeper cannot easily remove and replace them.
- In its primitive form, only one end of hive could be open, but in more advanced forms, each end of cylinder will be fitted with removable closure.
- The traditional backyard beekeeping practice exists to safeguard area for honeybees at homestead.
- The advantages of traditional backyard beekeeping practice could result in simple construction, no need of improved beekeeping equipment; no need of skilled manpower.

Types of traditional hives systems

1. Grass hive

- Dry grasses are woven together in basket or cylindrical form, with entry points at both ends.
- The hive is installed high in tree-tops to avoid termites.
- Many beekeepers lower it carefully at harvest time, while others drop it carelessly by cutting suspension rope.
- Due to weakness of material, hives usefulness will result less than one year, and can be used for seasonal beekeeping.

2. Gourd hive

- Such hive gives natural hollow for bees, which are too small for average bee colony, so that their use often induces swarming.
- Gourds are of two types where one is more or less pot-shaped while other has long neck which is attached to pot section.
- Both gourds are installed by suspension cord or by resting the mouth on wooden peg.
- Many gourd hives needs to be broken in pieces before honey and brood combs can be removed.

3. Log hive

- ❖ They are of two types.
- ❖ In certain isolated areas, ciba or palmyra palm produces natural hollows for interested beekeepers to use as hives.
- ❖ When plant dies, beekeeper waits for termites to consume soft inner pith, where tree gets felled and cut into pieces and ends are sealed with woven grass, whose ends provide entry and exit for bees.
- ❖ In other type of hive, tree is felled and cut into cylindrical logs which are carefully scooped out to form hollows. These are sealed, leaving some small holes for exit and entry.
- ❖ At harvest time, hive gets split open and honeycombs gets removed which further gets rejoined for bees to start with for honey crop.
- ❖ The log hive that are simple in construction is expensive and inefficient.

4. Barrel hive

- ❖ It uses metal and wooden barrels.
- ❖ These barrels contain pigs feet or wine, which were used as hives.
- ❖ Many of these old barrels are deteriorating, and no new ones are there to change them.

5. Clay-pot hive

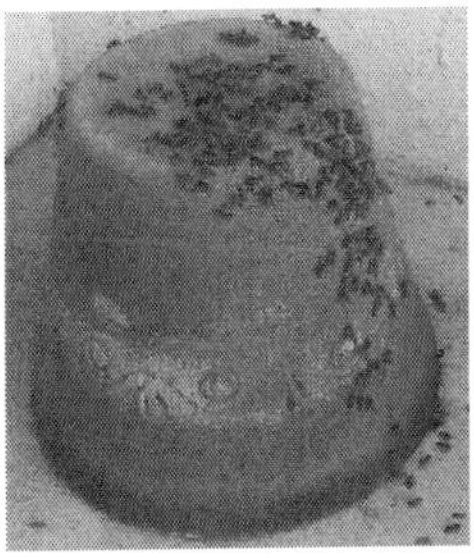

- It is the cheapest and most durable of among all traditional hives.
- It is very popular as pot is similar to type used to carry water or other liquids, and can be modified to provide wider mouth and small mid-section hole for exit and entry.
- Pots were made by elderly women, are bisque-fired, and inner part is smoked as part of baiting which are baited with cow dung or other waste and installed on ground or on pegs in trees.
- In many areas, pots are turned upside down directly on ground, for beekeepers to find, when they are installed on flat plate or wood, bees glue the plate firmly to hive with propolis, making harvesting tedious.

2.1.3 Modern hives system

- Modern beekeeping system obtain maximum honey crop, season after season, without harming bees.
- Modern movable frame hive consists of precisely made rectangular box hives which superimposed one above other in a tier.
- In this, number of boxes is varied seasonally as per population size of bees.
- The design of modern beehives is based on discovery made by many modern beekeepers, which noticed that when bees build their combs, they always leave exactly similar amount of space between them.
- On basis of finding, beekeepers invented hive with frames separated by bee space, in which the bees could build their comb.
- The frames are arranged such that they can be removed individually without disturbing other combs and crushing of bees, where sides and bottom of frame gives very good support for comb.
- Many communicating hive boxes can be stacked one above another, where queen can be confined to lowest or brood, chamber using queen excluder.
- With queen excluder, upper chambers can be reached by workers which contain only honey-comb that further makes hive inspection and many other management practices possible by turning art of beekeeping into full-scale industry.

Types of modern hives system

1. Top-bar hive

- In top-bar hive, the frames are replaced by simple modification of top bar of frame where bees build their combs by hanging down from centre of bar.

- As the combs are not supported on all four sides, they break more easily, but as they are fixed to top-bars and not to hive body, they can be removed and replaced any time for inspection or other management practices.
- Since boxes are build to receive top-bars, clay pots, barrels and many containers used by traditional beekeepers can easily be fitted with top-bars.
- There exists three types of top-bars which are V-shape or ridge, with groove and pointed starter which is first to be used by beekeeper who has no beeswax or comb foundation to serve as guide for bees.
- The problem with such hive is that new honeycombs can easily break away from the bar, so beekeeper should handle top-bars with combs carefully.
- Ridge system can be built quickly as compared to V-shape, but its manufacture needs use of electrically powered machincs for cutting groove in centre of bar.
- The beekeeper fixes strips of wax about 6 mm thick in groove to form guide for bees and made glued to wax foundation in groove to build their combs along it.

Advantages

- Such hive is cheaper and easier to produce, as any semi-skilled carpenter can make it using few simple carpentry tools.
- There is little or no need to import anything as all materials needed can be obtained locally.
- The hive can be opened easily and quickly as there is little or no need to for hive tool.
- The top-bars can be constructed to overlap the sides of hive body slightly, which makes it possible to use thumb to pry up the top-bar.
- Top-bars occupied by combs can easily be detected, so that hive gets opened from empty side to avoid crushing of bees between top-bars while lifting first comb.
- Bees in this hive can easily be controlled while harvesting or inspecting the combs. In this, smoker puffs smoke by opening created by removal of one top-bar.
- The top-bar hive is lighter to carry, even when the colony is inside.
- More beeswax can be produced so sales of beeswax increase beekeeper's earnings and solve great national problem as beeswax is a multi-purpose industrial raw material which is required by factories and craftsmen.
- There is no need to use queen excluder, which at the moment is not present. In practice, bees keep their brood chamber separate from honeycombs as clean honey can be taken away, leaving brood combs undisturbed.
- Honey production can be high, as well-managed hive with good strong colony can produce between 50kg-120 kg of honey annually.
- Honeycombs adulterated with pollen can be of high value as pollen is nutritious, where only nutrition is passed through honey harvested using combs.
- Only few extra top-bars need be held in stock to replace worn-out or damaged bars.

2. Groove top-bar

- ❖ This type of bars can be built quicker than other ones.
- ❖ Nevertheless, the manufacture of these hives requires the use of electrically powered machines as you can't cut the groove in the center of the bar.

3. Pointed starter

- ❖ It's also connected with the groove top-bar construction.
- ❖ Nevertheless, this type of hive can't be constructed by a carpenter.
- ❖ It also requires the production power.
- ❖ Nevertheless, the bees find it with ease to construct their hives inside such a box.
- ❖ The combs are always firmly attached.

2.2 Study tools/equipments required for beekeeping

Equipment required by the beekeeper

Smoker

- ❖ A smoker is most valuable tool for working bees.
- ❖ It calms the bees and reduces stinging.
- ❖ Many Beekeeper uses smoker to generate smoke from incomplete combustion of various fuels as smoke calm the bees and makes inspection of bees hive easily.

- It is manufactured using highest quality of stainless steel, with a hook on front to hang on hive.
- It has two main parts: container- which is a metallic can, big enough to carry enough dry material to last at least 40 minutes while other is bellows section- which puffs air in container to drive the smoke out of can.
- The container is loaded with wood shavings, smouldering cow-dung or any dry material which provides white smoke.
- The smoke renders bees docile, so that the beekeeper can work undisturbed.
- Pine straw, grass and burlap serves as good smoker fuel

Hive

- Hive with movable frames allow easy inspection and honey removal.
- Hive design is efficacious for other management practices and for bees.
- In this, inner dimensions and its parts are very precise which are based on certain dimension called as bee space which is 5/16-inch wide or deep.
- It needs proper spacing as gaps being too wide, allow the bees to build brace comb and glue down the movable frames.

Hive Parts

- Bee hive consists of following parts:
- Hive outer cover: It provides weather protection.
- Inner cover: It save bees from attaching comb to outer cover and provides insulating dead air space.
- Shallow honey supers: It has frames of comb in which bees store surplus honey which is harvested.
- Queen excluder: It is kept in between brood nest and honey supers which helps in keeping the queen in brood nest, so brood will not take plave in honey supers.

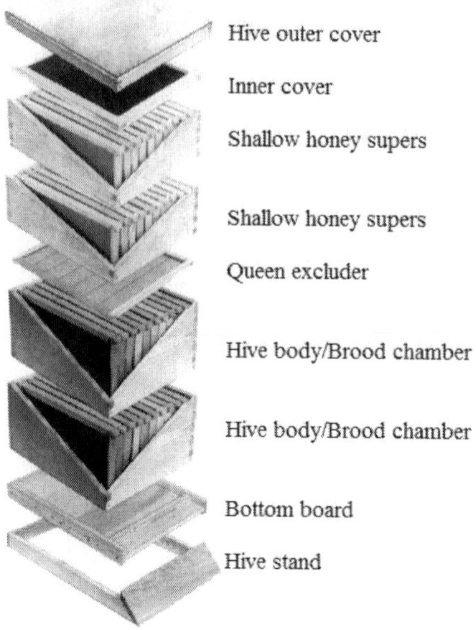

- ❖ Hive body or brood chamber: It is a big wooden box which holds 10 frames of comb. Such space is reserved for bees to rear brood and store honey for their own use. In this, one or two hive bodies can be used for brood nest. Two hive bodies are common in cold winter regions, so beekeepers in mild winters successfully uses one hive body.
- ❖ Bottom board: It is a wooden stand on which the hive rests. The set bottom board on bricks or concrete blocks keep it off the ground.
- ❖ Hive stand: It keeps hive off the ground so that it is less likely to rot, flood or attacked by termites. It can be made as simple as few bricks stacked under each hive corner, or can be of wooden frame with alighting board.

Frames and foundation

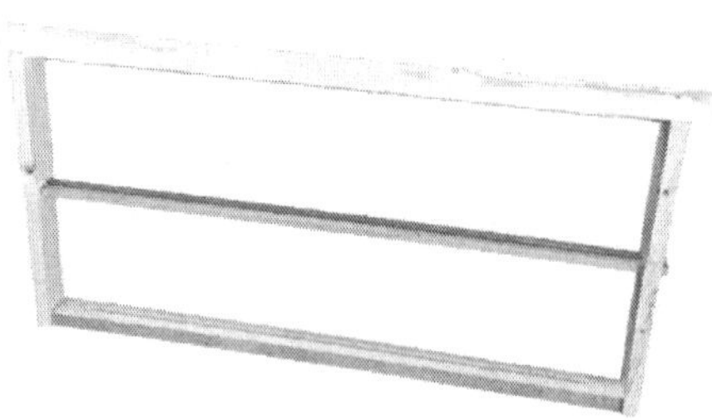

- ❖ Frames are inside parts which holds the comb.
- ❖ They consist of top, bottom and two end bars.
- ❖ The wide part of end bar is keeled on one edge.
- ❖ It is kept in the hive so that the keeled edge of one frame abuts the flat edge of other.
- ❖ It helps in keeping the comb-building regular and allow easy inspection and honey removal.

- Hives have frames of similar length, but there exits with different depths and styles.
- The wooden frames hold the sheets of beeswax or plastic foundation which are imprinted with shapes of hexagonal cells.
- Beekeeper fits the frame together in such a way that they keel on left end bar toward him while keel the right end bar away.
- If beekeeper rotate the frame, then keel will still toward him on left side and away from him on right.
- Beekeeper should uses many nails to fasten the frame together, otherwise, it can pull apart when comb is full with honey.

Knife

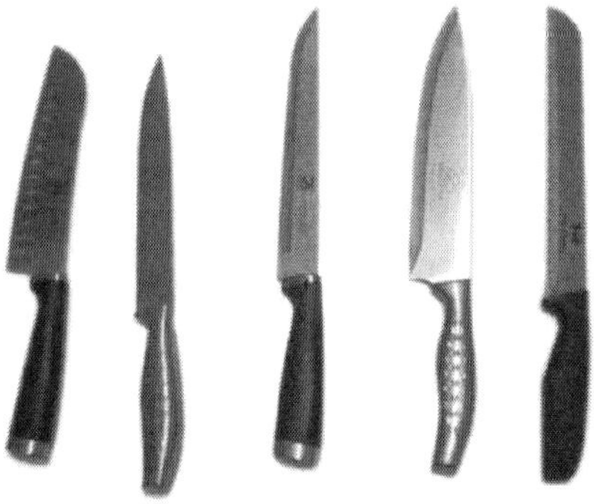

- It is required to pry open top-bars or frames which are normally glued to hive body by bees.
- It is a useful for cutting part of comb which is attached to hive body, that separates two combs that joins together, and cut out honeycomb from top-bar during honey harvest.
- It can do all functions of hive tool, but hive tool cannot be used to cut bee combs as neatly as is required.

Brush

- Brush or quill is used to brushed bees gently in container or hive.
- Brush with soft hairs is used for brushing of bees, but if beekeeper can easily obtain strong, large quill like an ostrich or turkey feather, in such case, there is no need to have a brush.
- The quill of big bird is better than any artificial device for brushing of bees.

Feeder

- ❖ If required pollen & nectar are not available from flowers, then beekeepers provide artificial feeding like sugar syrup to honey bees for keeping them alive, so feeder is used for artificial feeding.
- ❖ It can be a jam jar or special container which can be turned upside down and so arranged that water trickles slowly from it for bees to drink.

Queen Catcher

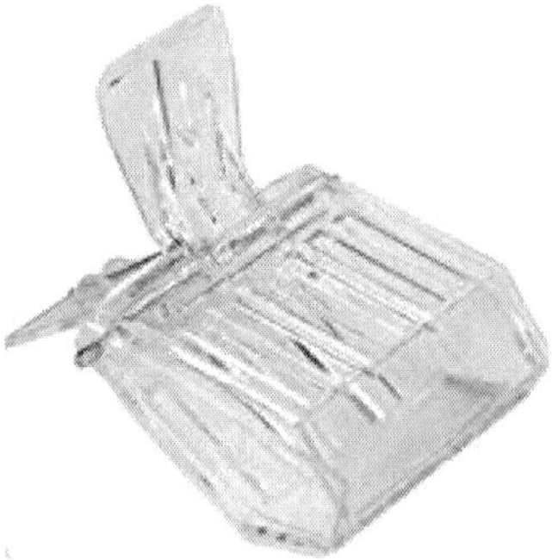

A queen catcher is a handy tool to have when you want to keep the queen separated for a while. A prime example of this is when you are going through your hives. Sometimes it is easier to place her in this catcher so you don't lose her in the process.

Pollen Trap

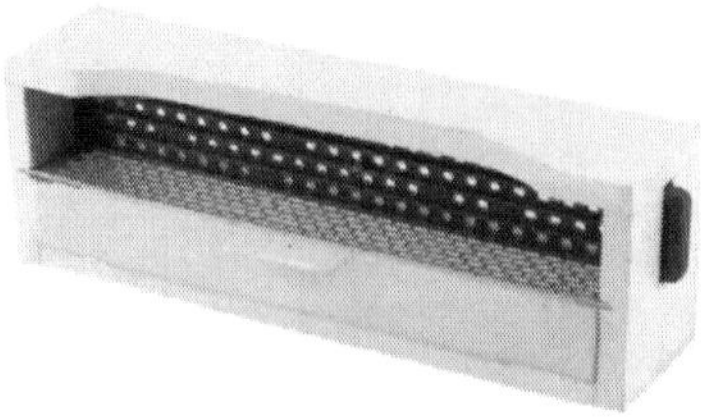

- ❖ It is used to collect pollen by installing it at the entrance of bee hive.
- ❖ There are varying designs for pollen traps.
- ❖ The pollen trap makes access to hive harder for foraging bees.
- ❖ In process of climbing by pollen trap wires, many pollen is loosened from bee's pollen basket and falls in collection container.

- ❖ Pollen collection works best in area with many pollen sources all through the year.
- ❖ Such traps vary in size, appearance and method of installation on hive.

Protective clothing

- ❖ Beekeeper is advised to get suitable protective clothes to keep bees from reaching the flesh, so bee suit, gloves, veil and pair of boots should be acquired before honey is harvested or any work involving opening of hive is undertaken.
- ❖ While working with bees during daylight hours, light-coloured clothing should be worn for night work, so beekeepers should wear dark colour cloothes.
- ❖ Bee suit cover all parts of the body except head, hands and feet and should be worn to harvest honey and control the brood nest during daylight hours.

Veil

- ❖ It is the most important as beekeeper can use such hat as netting is sewn firmly around it and attached at the back by piece of cloth.
- ❖ It protects the head, face and neck from attack.
- ❖ Wire veils keep bees farther away from face as compared to those, which are made of cloth.
- ❖ Black veiling is normally easier to see through as many beekeepers prefer to wear bee suit.

Hive tool

- A hive tool is most useful piece of beekeeping equipment.
- It can be used to pry up inner cover and pry apart frames, scrape and clean hive parts and involves in many other jobs.

Gloves

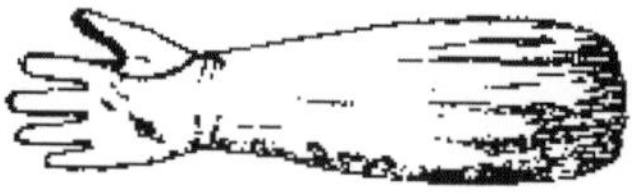

- The gloves used by beekeeper must be sewn with good, flexible white leather to protect the hand and fingers from stings and helps beekeeper to scoop up bees with hands if required.
- The bee stings on hand or fingers are most painful and beekeeper is urged to get gloves to ensure that he/she works with little or no difficulty.

Boots

- A pair of long boots is important to protect the feet from stings.
- When they are not available, pair of light shoes and thick white socks can be worn.
- Dark or black socks should only be worn at night when bees vision is poor.

Honey extractor

- It is a mechanical device which is used in honey harvest.
- A honey extractor extracts honey from honey comb without destroying the comb.
- It work by centrifugal force where a drum or container holds a frame basket which spins, flinging the honey out.
- In this, wax comb stays intact in the frame and can be reused by bees.
- The size of such frames ranges from 2-16 frames.
- The purpose of honey extractor is to collect honey from combs.

Queen excluder

- Queen excluder is a selective barrier in beehive which allows bee worker not larger queens and drones to traverse the barrier.
- It is either a sheet of perforated metal or plastic or wire grid in a frame.
- The openings are limited to 0.163 inches (4.1 mm).
- The idea of queen excluder is to limit the queen's access to honey supers.
- If queen lays eggs in honey supers and brood develops, it is difficult to harvest clean honey product, making fall management difficult.

Chapter 3

Beehive/Honey Bee Colony Management

3.1 Understand factors during site selection for successful establishment of apiary

3.1.1 Apiary

- An apiary is a place where beehives are kept.
- It can contain hundred or more bee colonies.
- It should contain only about 10 hives per km^2.

3.1.2 Factors considered in selecting of Apiary

Site selection

Selecting a good site for apiary can make huge difference to honeybee health, and hive's honey harvest. Poor sitting of apiary may lead to lower production in honey, high cost of production, conflicts with human and many others. It is, therefore, necessary that beekeeper should consider all possible factors that will enhance the life and activities of bees for higher honey production.

Following factors should be kept in mind while considering site for apiary which covers enquiry related to:

- Will the site causes nuisance to neighbors or general public?
- Is there any forage for honeybees or are there any apiaries nearby?
- Is the environment of site suitable for bees?
- Is there convenient access, with minimal carrying for beekeeper to bring equipment and remove honey supers?
- Is space suitable for number of hives?

Availability of flowers

- Site for apiary should be near a nectar source like flowering trees and shrubs, food crops and cash crops to attract bees.
- The beekeeper should know the flowering periods of various plants in and around apiary.

- In this, plants so selected should be produce high quality honey.
- The best beekeeping vegetation areas could be forest woodlands, grasslands with dense covers of flowering herbs/shrubs, thickets, agricultural crops yielding nectar in abundance can be good beekeeping sites e.g. sunflower, coffee, sisal estates legumes, bananas etc.

Availability of water

- Site for apiary should be close to perennial source of water.
- If there is no permanent source, in such case, beekeeper needs to provide water in neat colourful bowls or containers with floating sticks for bees to step on and avoid drowning.
- Water is necessity for bees for various uses in hive like cooling, feeding of larvae and own use.

Accessibility and topography

- Apiary must be accessible for easy management and transportation of honey.
- Flat or gentle sloppy lands are more preferred for apiary for easy management.
- Beekeeper should not select such site that entails climbing fences or crossing ditches to enter, but should select vehicular access right up to hives if necessary.

Shelter/Shade area

- Beehives should be sheltered against strong winds, direct sunlight and wind blowing off the top covers.
- The swinging of hives makes the bees rowdy and hostile.
- Beekeeper should provide shade by using trees/branches.

Fence/hedge

- Trees and bushes should surround the apiary as it makes the bees to fly high when leaving and returning to apiary which reduces risk of nuisance to nearby farms activities.
- The area should be fenced to exclude livestock and other animals that might disturb bees.

Proximity of human activities/conflicts

- ❖ Site of an apiary should be selected away from public places or areas where there are no human activities.
- ❖ Avoid areas close to schools, markets, hospitals, playgrounds, highways, cultivated fields etc. as bees can become nuisance to people.

Presence of pests and use of pesticides

- ❖ Apiary should be free from areas with frequent attacks by pests.
- ❖ The site so selected for apiary should far from fields where there results pesticides applications, to avoid bee poisoning and honey contamination.

Drainage

- ❖ Beekeeper should select the apiary site where there results proper drainage to avoid washing away during flood.
- ❖ Waterlogged soil cause rotting of hives and posts.

3.2 Study factors for proper installation of bee hives in apiary

3.2.1 Installation of beehives

- ❖ A hive can be placed in between two trees or from sturdy branches of big trees.
- ❖ It can also be installed on a platform or a rock.

Hanging beehives

1. It is cheaper to hang a beehive than to install it on platform.
2. The lizard which is an important hive predator, does not seem to pose serious danger.
3. Cattle and other grazing animals cannot tip the hive over.
4. Running water cannot carry the beehive away.
5. It is easier to save ants from reaching hive than when it is installed on stand.
6. A suspended hive can swing, so bees become alert and are prepared to pounce on beekeeper.
7. Honey-harvesting and brood-nest control are difficult to execute during the day.

8. It is not easy to change the location of hive, so while removing it from tree, the least false movement can result in tipping it over and jarring whole contents.

Installing hive on stand

1. It is easier to place the hive on stand and remove it.
2. It is easy to move hive and stand to another spot.
3. The beehive does not swing about even if beekeeper is working.
4. Honey collection and brood-nest control can easily be carried out even during the warmest time of day.
5. Grazing animals can knock the hive over.
6. Legs of stand can easily be used by lizards to reach the hive unless they are protected by lizard guards.
7. It is more expensive and tedious to make a reliable stand than to buy a metallic wire for hanging a hive.
8. Easy movement facilitates easy stealing.

3.2.2 Inspection of honey bee colonies and their record keeping

Inspection

- Visual inspection of colony is required which includes only representative sample of hives in apiary.
- All hives must be accessible and can't be double stacked, banded or presented in manner as to prevent inspection.
- Growers should provide location for beekeepers to place the bees to remain accessible during pollination period.
- Bee colony is complex organism of tens to thousands of bees.
- The sophisticated behaviors taking place inside beehive will not always be evident during the inspection as it is a gradual trend, rather than rapid change which provides clue to beekeeper that something is amiss, so many beekeepers keep track of bee happenings inside the hives. When beekeeper gains experience, there will be wealth of information in record which can add historical and environmental context to decisions that can be made.

Record Keeping

Keeping track of queen

If queen moves between the hives, then beekeeper should keep track of it, which should include details related to:

- age
- color
- original source

Keeping track of general information

For each inspection, beekeeper should keep records related to situational and environmental information which should cover information on:

Weight

- ❖ It is a common to record the weight of hive as it can vary considerably by ebb-and-flow of seasons.
- ❖ It is helpful in understanding the buildup of honey.
- ❖ Hive monitors can be used to collect data by smartphone which can be uploaded on Internet for analysis.

Brood counts

- ❖ At each inspection, beekeeper should estimate the brood counts for various phases.
- ❖ Estimation of brood count can be done quickly once beekeeper has idea about cells per inch, across the frame.

Tracking of eggs, uncapped larvae and capped brood

- ❖ It will uncover small changes in pace by which, queen is laying.
- ❖ If the queen is laying at reasonably constant pace, then ration of uncapped larvae to eggs is roughly around 2:1.
- ❖ If ratio of larvae to eggs decreases, then it shows that, queen has increases her egg-laying pace.
- ❖ Beekeeper should also track the number of queen cells as seen during inspection and drone cells.
- ❖ As bees are inclined to give 25-30% of brood as drones in wild which is more than what beekeepers needs.
- ❖ If a queen is lost, worker bees may lay eggs and without their ability to fertilize eggs, it results in drones.
- ❖ Tracking of drone cells will help to alert a beekeeper to queenlessness.

Brood pattern

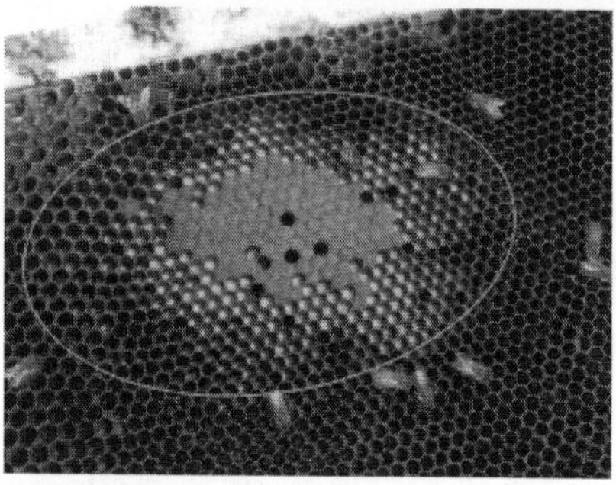

- The consistency of brood pattern is also required to be recorded for good beekeeping.
- A strong queen can create brood pattern with few gaps, as all queens leave at least some, whereas spotty brood pattern shows sign of weak queen.

Behaviors

- During inspection, beekeeper should record bee's mood.
- Bees react to all sorts of stimulus and is not always easy to locate direct cause of any unexpected behavior.
- Aggressive behavior from normally docile bees results as a clue, which beekeeper requires to investigate further, during spans multiple inspections.

Diseases

- All colonies experience disease or mites at certain point in their existence.
- Keeping accurate record of diseases can help tremendously.

3.2.3 Installation of bee hives

Instruments

To install the bee hive, beekeeper should use following equipments:

- Bee suit (Jacket, pants)
- Gloves
- Hive boxes
- Screened or solid bottom board
- Frames
- Hive tool
- Hive stand or cinder blocks
- In-hive or board man feeder
- Hive entrance reducer
- Spray bottle
- Sugar water (1:1 ratio)

- Mini marshmallows or candy plug
- Bee brush

Factors

Placement of hive depends on size of property, proximity to neighbors, current landscaping, access, etc. Beekeeper should see that bees are flexible, and it is best to choose location/ orientation which can minimize disturbances to neighbors, animals etc.

- Orient the entrance of hive to south or southeast location so that the hive receives 6-8 hours of direct sunlight in the morning.
- Ensure that area in front of hive should be clear for 10-15 feet, as such area allows clear flight path for bees, since they leave and return to hive. Beyond 10-15 feet, the bees' flight path should be above the average human height.
- Place your hive in an area safe from animals and people. If children are going to be near the hive, adding a small, inexpensive, wire fence around the hive may be desired.
- Beekeeper should see for low traffic area where bees are less likely to be disturbed.
- Use hive stand or cinder blocks to elevate the hive to 12-18 inches off of the ground to keep it safe from skunks and other animals which may want to disturb the hive.
- Place large rock on top or strap down the hive to keep it safe from raccoons, wind, etc.

Installation

Step 1: Beekeeper should prepare woodenware and remove two or more frames from box and put little water on screens with hand mister.

Step 2: Now lift the screen and dump in bees next to pry out staple that attaches screen to center post.

Step 3: Leave the top edge attached, and slice along bottom and side edges using matte knife. Beekeeper can also use pliers to pull the wood lath which holds screen on sides and bottom and should lift the screen.

Step 4: Now set it on top of top bars where the screen to face towards is loose on three sides but to be attached on top.

Step 5: Using beekeeping gloves, gently lift the screen out of the way.

Step 6: Lift the screen away and tip the mass of honey bees in box and shake.

Step 7: Quickly slide in all the frames as bees will start climbing on frames and box walls.

Step 8: Now place the queen cage by temporarily placing outer cover on catty corner to shade bees while preparing installation.

Step 9: Pull out the syrup can by using strip by suspending the strip in queen cage.

Step 10: Remove the queen cage and see that some bees will be hang on the screen where queen along with others are inside the cage. Gently blow on bees to clear them off.

3.3 Study important plants/flora with source of nectar or pollen

3.3.1 Introduction

- Honey is a sweet viscid fluid collected from nectaries of flowers elaborated for food by many kinds of insects.
- Honey is a nectar which is obtained from flowers by worker bees, which on modification in stomach of bees, gets stored in cells of comb for nourishment of young brood.
- Production of honey and other products depend on availability of floral resources which is important for beekeepers.
- Honeybees depend on flowering plants for their food requirements.
- Nectar is sweet liquid which comes from floral and extra floral resources that serves as raw material for honey while pollen is highly proteinaceous food for bees.
- The plants that yield both these substances are referred as bee pastures.
- Honey is only medium where no bacteria can germinate due to presence of sugar content and due to its acidic reaction.
- Nectar and pollen plants are basic needs for beekeeping and honey production, their knowledge is required for beekeepers.
- Bee forage plants can be fruits, vegetables, ornamental plants, crops, medicinal plants, herbs, shrubs, bushes, trees, forest and weeds.
- Honey bee gather nectar and pollen from bee plant as their food and pollen foraging activities of honey bees are varied time to time.
- Pollen gathering activity is dependent on the availability of pollen yielding flowers and environmental conditions such as sunrise, sunset and day temperature etc.
- The interaction of flora determines pattern of pollen gathering activity.
- As honey bee doesn't visit all plants for nectar or pollen, so identification of such plants which supply resources, plant communities and phonological relationship between honey bee plants and honey bees are important for practical beekeeping in assessing potential of area bee keeping.
- Good beekeeping area shows nectar and pollen plants in abundant with relative long blooming season.
- In honey flow season, there are around 29 bee plants, out of which, 9 plants did not provide pollen and 5 plants did not provide nectar.
- In dearth period, out of 34 bee plants, 15 plants did not provide pollen while 12 plants did not supply nectar.
- In extended period, out of 4 bee plants, many of them provide only pollen and also 5 bee plants that gives pollen and nectar all through the year.

3.3.2 Flowers which attract bees include:

❖ Blackeyed Susan: The flowering period is summer blooming flowers.

❖ Saffron flower: The bloom period is October and November.

❖ Jamun flower: It is a fruit flower which blossoms in cold areas during May and June.

❖ Sheesham: It is an evergreen rosewood tree that flowers from March to May months.

❖ Mango: The flowering season start from December till March.

❖ Ber: It blossoms in cold and hot region during September.

❖ Aster: It blooms when other flowers fade which gives honeybee required pollination. It blooms in late summer.

❖ Lemon balm: It is a nectar rich flower that fed honeybees.

- Purple coneflower: It is a flower that serves as honeybee magnet which gives pollen and nectar to foraging bees.

- Snapdragon: It release strong scent which attract bees.

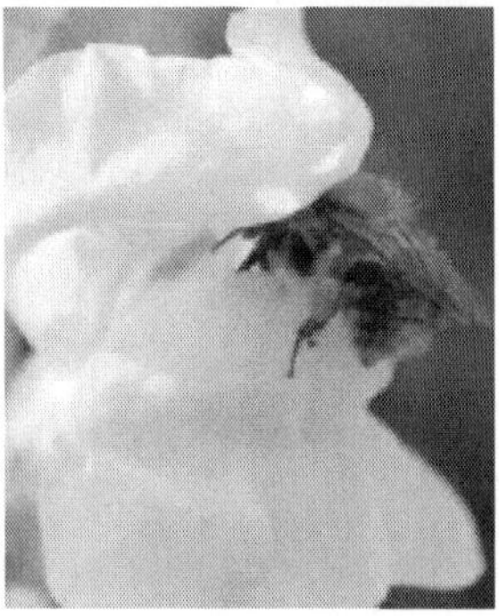

- Golden rod: It blooms from August to November.

- Butterfly Bush: It serves as perennials bush/small tree, and needs pruning to promote profuse flowering.

- Sunflowers: It has large central discs that give opportunity for bees to forage for nectar and pollen.

- Salvia: It gives protein-rich pollen for bees in early spring.

- Blanket flower: It is a daisy-like flower which gives sustenance for bees. It blooms from May to October.

3.3.3 Plants/flora of nectar/pollen

Nectar rich plants

- Phaseolus vulgaris
- Cyamopsis tetragonolobus
- Momordica charantia
- Mangifera indica
- Pongamia pinnata

- Jasminum sp. Gladiolus communis
- Echinops echinatus
- Santalum album

Plants visited by bees for nectar:

- Tamarind(Tamarindus indicus)
- Neem(Azadirachta indica)
- Soapnut tree(Sapindus spp.,)
- Eucalyptus spp..
- Pungam(Pongamia pinnata)
- Moringa tinctoria
- Prosopis spicigera

Pollen rich plants

- Solanum melongena
- Lycopercicon esculentum
- Capsicum sp
- Cucumis sativus
- Cucumis melo
- Citrullus lanatus
- Amaranthus gracilus
- Cocos nucifera
- Eugenia jambolana
- Justica adhatoda
- Melia azedarachta
- Cassia tora

Plants that supply pollen to bees:

- Sorghum
- Maize
- Roses
- Finger millet
- Bajra
- Castor
- Tobacco

3.4 Study factors affecting nectar and pollen production in bee flora

Introduction

- Nectar and pollen production potential of plants varies due to their taxonomy, geographic distribution, age, season etc.
- Floral characteristics such as flowering time, length of flowering, flower density, color of petals, amount, location and nutritional status of nectar and pollen influences plant to honey bees.
- Bees can distinguish green, blue, white and violet coloured flowers but insensitive to red flowers.
- Bees prefer flowers secreting nectar with higher sugar content.
- Floral nectar plays an important role in plant reproductive success by attracting pollinators.
- Nectar traits of flower depends directly on plant characteristics apart from environmental factors and shows extensive flower and plant level variations.
- The suitability of flora depends on liking and disliking of nectar by honey bees.
- The availability of bee flora in location shows potential of area for honey production.
- Suitable bee flora with well distribution of flower availability over time is required for rearing of healthy bees.

3.5 Seasonal management of honey bee colonies

Whether you keep bees for profit or as a hobby, keep records to manage your hives efficiently. They allow you to see if your effort is paying off.

One type of record is a diary in a weatherproof container attached to each hive. In the diary, record: date the colony was established, dates of inspection, date it was requeened, honey yield, occurrence of disease if any, and other facts which help you manage a particular hive.

Record Keeping

Nectar Flow Records for 19__			
Plant	**1st Bloom**	**Foraging Dates**	**Last Bloom**
White Clover	15 May	5 Jun to 15 Jul	3 Aug
Sourwood	25 May	4 Jul to 28 Sep	5 Aug
Boneset	20 Jul	25 Aug to 15 Sepl	15 Oct
Nectar Flow Records for White Clover			
Year	**1st Bloom**	**Foraging Dates**	**Last Bloom**
19__	15 May	5 Jun to 15 Jul	3 Aug
19__	17 May	4 Jun to 20 Jul	4 Aug

- The first style shows the succession of bloom more clearly, while the second style shows the range of dates for which you can expect nectar flow from a particular plant.

Seasonal Management

- Depending on suitability of climatic conditions and availability of plenty of bee flora, when climate and floral condition are suitable, proper seasonal management of honey bee colonies serves as success of beekeeping.
- For manipulation of honey bee colonies, beekeeper must understand the behavior and needs of honey bee colonies and should follow required operations to fulfill their needs to have maximum efficiency of bee colonies and benefits from beekeeping.
- The bee management operations vary in time and space and with species of honey bees to adopt for beekeeping.

The seasonal management operations to be undertaken in honey bee colonies are based on certain principles such as:

- Beekeeper should help bees in building up colony strength in 1-1 ½ month before major honey flow season and not during honey flow, so as to harvest bumper honey crop by changing old exhausted queen bees, every year in colonies with young mated/ prolific queen bees and should ensure required space by adding worker combs/comb-foundation sheets and new supers and uniting unproductive week/laying strong worker colonies.
- Beekeeper should maintain optimum strength during dearth season so that bee colony does not perish but can perpetuate to ensuing good season.
- Bee colonies should not suffer for want of space for brood rearing, honey/pollen storage and honey ripening during active season.
- Beekeeper should make provision of space in colonies, to lower the congestion and should check swarming instinct in honey bee colonies.
- Beekeeper should extract only surplus and sealed honey at right time around 7-10 days before cessation of honey flow and should manage colonies to check robbing, in apiary, if required.
- Beekeeper should help the colonies suitably to tide over adverse climatic conditions, floral dearth periods by migration of colonies to suitable localities.
- Beekeeper should manage the incidence of bee enemies and diseases cautiously.

Spring Management

- Spring management varies from locality to locality and year to year as per weather conditions and availability of nectar sources in area.
- Beekeeper should modify following sequence of spring management practices for certain areas which includes:
- Early Spring (late March to early April)
- Tilt each hive back and clear bottom board of dead bees and debris.
- When finished, change the entrance reducer.
- Beekeeper should check for good egg and brood pattern which shows status of queen, where queen begun in late February or early March.

- If queen is missing or not laying in good pattern, beekeeper should plan to requeen in April when some nectar is coming into hive.
- Beekeeper should check for sufficient stores as supply should never be less than 10 pounds as bees can exhaust this in a week if no nectar comes in.

Mid-Spring (fruit bloom period)

- Requeen the hives as per the plans.
- Continue to watch for poor queens in other hives like spotty or sparse egg laying and requeen immediately if needed.
- Beekeeper should provide free access to hive entrance by removing entrance reducer and rank the plant growth in front of entrance.
- Wih progress of fruit bloom period, brood rearing move up in overwintering super and if such happens, then reverse the hive body and super so that super will be on the bottom.
- If full frames of honey and pollen are next to brood nest, then block their expansion and rearrange the frames in brood chamber so that empty frame can separates the brood from food stores.
- Since the frames are filled with brood, beekeeper should provide more space in a likely manner.

Late Spring (end of fruit bloom to start of main nectar flow)

- Beekeeper should see for signs of swarming and prepare empty equipment to hive swarms to increase the numbers of colonies or can provide expansion room.
- Just before the main nectar flow, beekeeper should rearrange the hive body and super so that super should again be on top.
- Start spurring for honey production and see section on spurring for instructions.

Fall Management

- Fall management prepare the hives for winter which requires following:
  - large supply of healthy young bees
  - abundant supply of food stores
  - sufficient protection from cold and disturbances
- A large cluster of bees is required to produce and conserve enough heat to survive. If the cluster shrinks in size during winter, then many bees die of old age or disease and then rest of colony will also die due to freezing.
- Since honey is the fuel, so bees use such fuel to generate heat, and hence an abundant supply is required for winter.
- Beekeeper should protect the hives from winds or bees will use up their honey too fast and might not be able to maintain living temperature.

Fall management practices will help bees in meeting satisfactory wintering conditions which involves:

- Checking the cluster having 5 to 12 pounds of bees or enough to cover 5 to 12 frames.
- Unite small colonies with stronger ones ands leave the extra hive body above the hole in inner cover to keep wax moths from destroying the comb.
- Feed the colony fumagillin-treated syrup if Nosema disease is in your area as such disease can reduce cluster size as it will die due to freezing.
- During late September and October, if colony does not have full hive body and shallow super of honey or about 50 pounds, then beekeeper should feed sugar syrup for shortage.
- Beekeeper should build windbreak or move the hives to protect hedge, wall, building, or some natural windbreak.
- Wrapping hive with insulation is not recommended as bees winter as well without it.
- Before initial frost, beekeeper should insert the entrance reducer as entrance reducer at this time will help in preventing field mice from entering the hive.
- Treat for varroa and tracheal mites by using vegetable shortening or menthol treatments for tracheal mites and Apistan strips for varroa mites.

Winter Management

- The main thing to check in winter is that the hives are not damaged by animals or storms.
- If beekeeper provides proper conditions in the fall, there is no reason to open the hive in winter as unnecessary opening will create problems.
- Unnecessary opening of hive can cause hive temperature to drop and temperature inside cluster to rise which stimulates brood rearing that deplete food stores.
- Cracks created by moving hive parts will not be resealed so hive will lose some of its insulating ability.

Seasonal Variance of Honeybees

- Beekeeping is based on respect for bees which must be approached with mood of observation and nurture rather than expectation and exploitation.
- Success comes from knowledge and understanding of life of bees and their needs.
- As bees are responsive to environmental influences, it is important to gain knowledge of local seasons, weather and flora.

January/February

- It is a quiet time period for bees.
- Ensure there is ample food in the hives to tide the bees over prolonged wet weather.

March/April/May

It is a more quiet time for bees in which beekeeper:

- Check hives for disease.
- Bees fill honey frames to insulate the brood.

- Ensure bees are working all the frames and should remove or add supers as necessary.
- Ensure hives are in warm sunny position and if not, then he/she should move them.

June/July

- Inspect hives sparingly - leave brood boxes alone unless problems are suspected.
- Inland areas have few flowers, so leave honey in hive.
- Do not open the hive in cold or windy weather.
- Be alert to possible early honey flows, e.g. Blue Gum.

August

By mid-august, queen expands her egg-laying to boost the population after winter. It is a time to manipulate all frames in the box, so in last 2 weeks of the month, beekeeper should inspect every frame in hive and should focus on:

- Check the general health of brood and look for any disease in it.
- Check that Queen is performing/laying eggs and larvae.
- Remove honey frames and old frames from brood box leaving only two honey/pollen frames on outsides, allowing 8 frames for Queen to lay.
- Check / fill / renew beetle traps.
- Extract surplus winter honey.
- Reject / replace any old dark honey frames.
- Do not open hives on cold windy days as hive will take 4 days to return to normal temperature inside the hive.

September

It's a busy time in the apiary where beekeeper:

- Inspect the hives which was not already manipulated since winter.
- Add new frames and supers to strong hives.
- Extract surplus honey and start collecting stand-out honey samples for show.
- Be prepared for swarming.
- Plan for splitting hives.
- When extracting should consider the 1/3 rule: 1/3 for brood, 1/3 for honey / 1/3 empty.

October

It's still spring, which continues as for September where beekeeper:

- Check for Queen cells.
- Add foundation to discourage swarming.
- The prime month for dividing strong hives and forming nucs.
- Increase vigilance for Small Hive Beetle.

December

In this month, honey flows which may ease off and increase rainfall where beekeeper:

- ❖ Leave stores in hives to tide the bees over prolonged rain events.
- ❖ Continue Spring management with emphasis on Small Hive Beetle management.
- ❖ Consider as the last month to form nucleus hives.
- ❖ Only strong hives may be split.

3.6 Migration of honey bee colonies

- ❖ In case of temperate bees, migration strategies are not used, as it is fatal to colony which migrates in entirety that collect enough stores to survive lengthy dearth periods, that encounters.
- ❖ Long-distance seasonal migration among discrete sites is common among vertebrates and rare among invertebrates.
- ❖ Many evidence shows that seasonal migration takes place in populations of tropical honey bees. In social insect, members of colony collectively choose migratory route to follow.
- ❖ Apis dorsata and Apis laboriosa are bees that have best migration behaviour having planned movement of all colonies which are predetermined alternative migration site.
- ❖ Asian honey bees do not usually store great amounts of honey so their survival strategy makes it to put effort in developing reproductive swarms rather than storing surplus food stuff that may be taken at any time by predators, leaving bees vulnerable to starvation in case of prolonged shortage of nectar or pollen.
- ❖ The response to diminish resources is to move to an area where food is more abundant. In many cases, open nesting bee species migrate twice each year.
- ❖ It is seen that common seasonal pattern for colonies arrive at end of wet season, where combs are built and period of rapid colony growth and strongest colonies produce reproductive swarms. By end of dry season, pollen availability is reduced, brood rearing diminishes and adult population declines.
- ❖ During a period, combs get attacked by wax moths and predators while parasitic mite populations will be at maximum where colonies move from site and start long migration to new locality.
- ❖ The travel of migrated colonies is not clear, but the travel settles in trees for rest periods, the workers will forage for stores as quiescent colony have minimal energy needs.
- ❖ The colony only moves again when it gather sufficient stores for next part of journey where movement is preceded by waggle dances on surface of colony that shows direction of movement.
- ❖ It is not known whether parts of dance refer to whole journey of just part to be considered on the day, when colony moves upto 20kms per day.

- It is not understood, how bees find their intended new nest site as no workers who travel have done the journey earlier whereas the queen passes such information.
- The probability is that bees follow an environmentally beneficial trajectory with sufficient forage and optimum temperatures.

Chapter 4

Manage Insects, Diseases and Nuisances in Beehive/Honey Bee Colony

4.1 Study different types of insect pests infesting honey bee colonies and their management.

Small Hive Beetles

- The small hive beetle is honey bee pest that appeared in late 1990s.
- The adult beetles are black or dark brown, and grows to about one-half centimeter.
- The larvae are white and is twice the length of adults.
- During pupa stage, small hive beetles burrow into soil surrounding the hive before emerging as adults.
- Adult females breed and lay eggs in cracks or small spaces in hive to complete the lifecycle.
- It is larvae of small hive beetle that do damage to hives which eats through combs as they feast on honey and pollen.
- In their wake, combs and caps are destroyed, and larvae's feces causes honey to ferment and discolor, which lead to completely abandon of hive in colonies.
- Small hive beetles are problematic for weakened or underpopulated hives.

Wax Moths

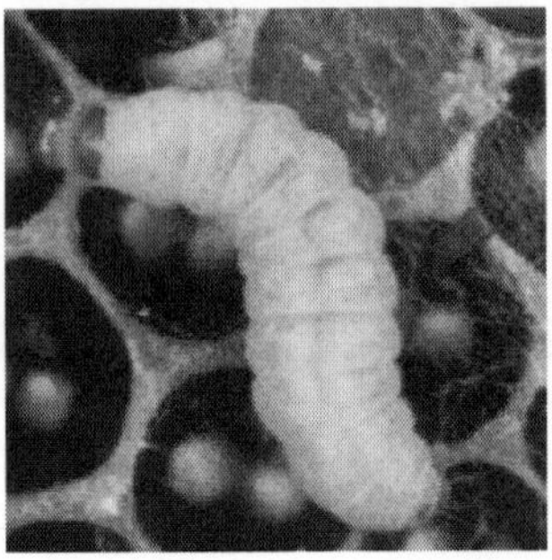

- The larva of wax moth are off-white caterpillars having brownish-black feet and heads.
- They are considered pests by beekeepers, but are commercially raised as food for insect-eating pets like lizards and birds.
- The larvae, waxworms are sold as bait to anglers, and are investigated for their ability to digest polyethylene plastics.
- Wax moths and waxworms do not directly attack honey bees, but their destructive behavior causes damage to comb and stored honey which leads to death of bee larvae in hive.
- Wax moths lay their eggs in hives as proteins found in can be used as brood comb which are vital for development of waxworms into adult moths.
- The waxworms chew through combs leaving frass, cocoons, spun tunnels and destroyed comb in their wake.
- Healthy colonies do not have problems controlling waxworms in hive, if weakened by disease or other issues affects colony's population.
- Beekeeper should store honey supers, as supers with dark comb or stored pollen are susceptible to waxworm damage in warm climates or temperature-controlled buildings.
- Wax moths and their larvae are easily killed by freezing temperatures.
- In warmer climates and warm storage facilities, para-dichlorobenzene products like Para-Moth may be used to control infestations.

Varroa Mites

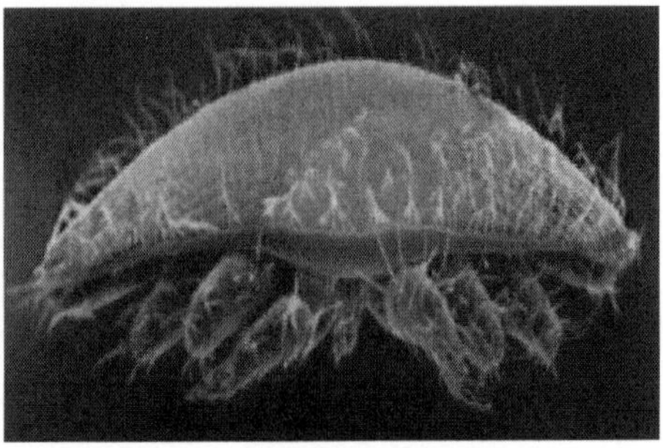

- Varroa mites are tiny, oval-shaped arthropods that are flat having brown appearance.
- These mites infest hives and feed on bodily fluids of bees of all ages like larval, pupal and adult stages.
- Healthy hives with ample population of bees are able to withstand the effects of moderate varroa mite infestations as they reproduce in brood, big, healthy colonies also rear lots of mites.
- Later in the season, during slow population growth or reduced hive vitality, these mites become serious problem.
- As such mites are known carriers of many viruses that make bees sick, these infestations often leave honey bee colonies weakened and more susceptible to disease.
- Colonies infested with these mites show developmental abnormalities like wing deformities.
- Chemical treatments like thymol-based Apiguard and amitraz-based Apivar, have proven safe and effective treatments for reducing mite populations in honey bee hives.
- Mechanical measures are effective at helping to reduce mite levels in combination with chemical treatments.
- Using screen bottom board in hives allows mites to dislodged by grooming to fall out of hive and away from colony.
- The Bee Gym was designed to help such process by encouraging more effective grooming habits by bees.

Tracheal mite

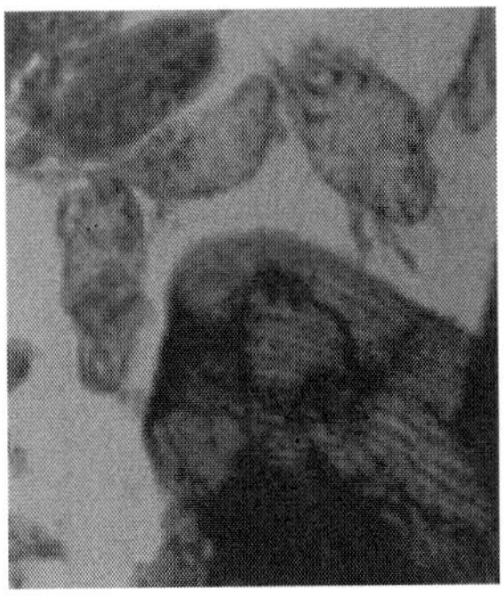

- It is a mite that infests honey bees.
- It is an internal parasitic mite which lives in tracheae, or breathing tubes, in thorax of adult honey bees.
- Tracheal mites are found in air sacs in thorax, abdomen and head.
- These mites pierce the breathing tube walls with their mouth parts and feed on hemolymph, or blood of bees

Bee louse

- ❖ Braula coeca, commonly known as bee louse is a wingless fly.
- ❖ The adults of the fly are small and reddish-brown in color.
- ❖ As many adult flies live on queen, but only one will be found on worker.
- ❖ These pests apparently do little harm.

Ants

- ❖ Ants are not serious pests in honey bee colonies, but certain species of enter in bee colonies to search for food or establish nesting sites.
- ❖ They are typically found between the inner and outer covers of hive and in pollen traps.
- ❖ As ants seldom disturb the bees, so they can be nuisance to beekeeper.

Mice

- ❖ Mice are serious pest of stored combs that inhabit active honey bee colonies during fall and winter months.
- ❖ Such rodents chew combs and frames to make room for building their nests.
- ❖ They urinate on combs and frames, making bees reluctant to use combs or clean out such nests in spring.

Skunks

- Skunks are serious threat to successful beekeeping as they hamper the development of strong colonies.
- They are insectivorous and will raid bee yards nightly by scratching on hive entrances and consumes large numbers of bees.
- Since their attacks are common in spring seasons, they occurs all through the summer and fall.

4.2 Study different types of diseases infecting honey bee colonies and their management

4.2.1 Introduction

- Similar to other living creatures, honeybee suffers from diseases. Across the world, research is underway for means of combating or preventing, but in certain areas, there results not much research on bee diseases. As seen, some of bee diseases found in temperate and sub-tropical regions of other parts of world may be present on the continent.
- The life cycle of honeybee starts with egg, which hatches in three days and then passes through larval and pupal stages before emerging as imago or young bee. During the brood stage, the insect gets attacked by bacterial, viral or fungal diseases.
- Beginner beekeeper cannot detect any brood disease unless he knows what a healthy brood comb looks like as it looks clean, which can be black, brown or white. A good, healthy queens can lay their eggs in clean cells, so beekeeper must watch the laying pattern of queen which can usually takes the form of concentric circles. During laying of eggs, eggs are laid at the centre of comb and then outward in rings to comb edges. The capping of pupae follows the same pattern from centre to edges. The regularity of brood in cells should be noted as good brood comb cells usually compactly filled by 5th and 6th days before sealing takes place. An irregular brood comb gives signal brood disease, so care should be taken, as irregularity may result of brood emerging.
- Beekeeper should watch a healthy larva carefully, as it coils like comma in a cell which is fleshy, glistening, juicy and white in appearance and cannot move from place to place in cell. The larva does not look brown, black or any other colour except white and if misshapen or found dead, then pupae must remain capped with seal and should not be punctured or sunken. With such irregularities, something will go wrong which may cause disease. Some diseases are serious and can wipe out entire colony as they gradually spread in other colonies and destroys whole apiary.

4.2.2 Types of Diseases

Bacterial Diseases

American foul brood (AFB)

- It is the most serious of all brood diseases which is followed by European foul brood.
- They are so called not because they are peculiar to each of continents but because the

two diseases were studied separately by American and European researchers at the same time.

Cause

- The disease is caused by spore-forming bacterium, a Paenibacillus larva, which only affects bee brood, where adult bees are safe from infection.
- At initial stage of colony infection, only few dead older larvae or pupae were observed, but subsequently, if remedial action is not taken, disease will spread in colony and can quickly spread to other colonies in apiary as a result of robbing, drifting workers, or contamination caused by the beekeeper's hive manipulations.
- The spores enter the bee colonies by foreign honey which is highly contaminated, so beekeeper should see near honey processing enterprises and waste disposal sites for disease spread.

Symptoms

- In initial stage of infection, isolated capped cells from which brood has not emerged can be seen on the comb.
- The caps of dead brood cells are darker than the caps of healthy cells, sunken, and often punctured while caps of healthy brood cells are protruding and fully closed.
- When disease spreads in colony, scattered, irregular pattern of sealed and unsealed brood cells can be distinguished from normal, compact pattern of healthy brood cells observed in healthy colonies.
- The bee brood affected by disease is usually at stage of older sealed larvae or young pupae, upright in cells.
- A protruding tongue can be found with rest of body that is already decayed where initially the dead brood is dull white in colour which changes to light brown, coffee brown and finally dark brown or black.
- Once dead brood have dried in scales, test cannot be used as dry brood lies flat on lower side of cell wall, adhering closely to it in contrast to sacbrood which is usually black or dark brown and brittle.
- A fine, threadlike proboscis or tongue of dead pupa can be seen protruding from scale, angling toward the upper cell wall.

Infection

The infection of such disease is spread by spores which are found in infected hive material, especially in combs. The contamination of larvae takes place by food containing spores which are administered to worker bees. The worker bees in attempt to clean cells, occupies by dead brood contaminated with disease and gets infected all over their body with sticky residues, and serve as carriers of spores to larvae.

As the colony is very sick and weak, it can be sacked by stronger colonies which serves as one of the ways by which the disease can be spread from hive to hive and apiary to apiary. Many times beekeeper with normal management practices promotes the spread of such disease whose risk factors can be:

- empty hives left after pollination services or migratory beekeeping;
- non-renewal of brood combs in the nest or non-replacement of the queens;
- feeding bees with infected honey or use of infected honey in supplement food;
- the queens could be more or less susceptible to the disease, according to their genetic characteristics;
- transfer of brood frames from one hive to another;
- the spores can be transmitted through the use of contaminated tools;
- purchase of swarms of uncertain origin.

Control

It is important for beekeeper to take necessary actions as soon as possible to control such disease which includes:

- destruction by incineration of infected colonies which can be in case of advanced stage of disease, weak colonies or low prevalence of disease in apiary;
- accurate disinfection of objects used for manipulation of infected hives covering equipments used;
- treatment of infected colonies as the treatments against disease are on-trade and medicines can able to guarantee about disinfection of hive from P. larvae.
- shook swarm method by shaking hives from infected combs in clean hive with new foundation.

Prevention

Proper treatments should be given by beekeeper to prevent the outbreak of such disease which covers:

- avoiding feeding bees with honey;
- periodic replacement of combs and queens;
- regular cleaning and maintenance of hives to prevent looting;
- renewing of combs after every 2-3 years which helps in prevention not only against the disease, but also for other bee diseases;
- replacing queen after every two years to guarantee strong colonies with large number of workers involved in removal of diseased brood and dead larvae.

European foul brood (EFB)

- Bacterium Melissococcus pluton is the main causative agent, but larva's death is accelerated by presence of Bacterium eurydice and others.
- The young larva is infected by taking in food containing the bacteria, which multiply in its gut where larva dies on 4th day, and worker bees leaves the cell with dead larva uncapped.
- Many times, infected larva does not die till it is sealed, which results in sunken and perforated cappings.

Cause

- The pathogenic bacterium of disease is Mellissococcus pluton.
- It is lanceolate in shape and occurs singly, in chains of varying lengths, or in clusters.
- The bacterium is Gram-positive and does not form spores.
- While many strains of M. pluton are known, all are closely related.

Symptoms

- Honey bee larvae killed by disease are younger which killed by American foul brood.
- The diseased larvae die when they are 4-5 days old, or in coiled stage.
- The colour of larva changes as it decays from shiny white to pale yellow and then to brown.
- When dry, scales of larvae killed by European foul brood, in contrast to American foul brood scales and do not adhere to cell walls and can be removed with ease.
- The texture of scales is rubbery rather than brittle, as with American foul brood.
- A sour odour can be detected from decayed larvae.
- The characteristic of European foul brood is that, many of affected larvae dies before their cells are capped.
- The sick larvae appear somewhat displaced in the cells.
- When scattered pattern of sealed and unsealed brood is observed in diseased colony, it shows that colony has reached a serious stage of infection and may be significantly weakened.
- European foul brood is transferred in same way as American foul brood.
- Melissococcus pluton as permanent form, does not form spores but capsules which are less resistant than the spores of P. larvae.
- The detection of M.pluton is carried out microbiologically where selective culture media are most appropriately used.
- The gene technological test is sensitive and is less suitable for detection of M. pluton in suspicious brood.
- A single-use test set is available based on serological proof like American foul brood test set.

Condition:

- Larvae reared in unfavourable conditions are susceptible to such disease as compared to those which reared in favourable conditions.
- The disease can be seasonal and occurs during and immediately after seasonal rainfall, thereby gradually diminishing till population of colony rises again in month of October where honey yield of affected colony drops.
- Such drugs as streptomycin, penicillin and terramycin control the disease.
- Immediately after disappearance of disease, queen gets removed and colony requeened.

Control

Beekeeper should take appropriate actions to control the infection of such disease by:

- ❖ destructing it through incineration of infected colonies during advance stage of disease, weak colonies or low prevalence in apiary;
- ❖ accurate disinfection of objects used for manipulation of infected hives, including equipment for operations;
- ❖ using shook swarm method consisting in shaking hives from infected combs in clean hive with new foundation.

Prevention

- ❖ ensure colonies have always available stocks of food (pollen and honey);
- ❖ do not use honeys suspected of being infected to feed the bees;
- ❖ do not move combs from a hive to another without checking their healthy conditions;
- ❖ renew the combs every 2-3 years (about 30 percent of the combs per year);
- ❖ remove the queen from the infected colonies.

Fungal Diseases

Chalk brood

- ❖ Chalk brood is derived from chalky appearance of dead brood.
- ❖ It is a fungal disease which is caused by Ascophaera apis, which causes serious problems to bee colonies in humid areas.
- ❖ Spores of fungus are ingested in brood food where spores germinate in gut, and growth of fungus causes death of brood, that takes place in pre-pupal stage.
- ❖ In Asia, chalkbrood is rarely considered to be serious honey bee disease, as in Japan, the disease has been reported to cause problems to beekeepers.
- ❖ In temperate America and Europe, cases have occurred where the disease has caused serious damage to beekeeping, so beekeepers in that areas are aware of such problem.

Cause

- ❖ The disease is caused by fungus Ascosphaera apis.
- ❖ It affects honey bee brood.
- ❖ The fungus forms spores during sexual reproduction.
- ❖ Infection by spores of fungus is seen in larvae that is 3-4 days old.
- ❖ The spores are absorbed either by food or by body surface.

Symptoms and Diagnose

- In this, larvae affect the different life stages on third or fourth day of larval life.
- The larva then die in first two days after capping, so bees must uncap the cells to remove the dead larvae.
- The disease produces mummification and/or calcification of larvae where larvae appear soft, assuming the hexagonal shape of cell and then dries out and become hard.
- The larger part of affected larvae appears white, but some become grey or black and may not present any symptoms if infection is less than 12%.
- The presence of little stones on bottom or at entrance of hive is typical.

Control

- As with other brood diseases, bees remove infested brood with their hygiene behaviour which is effective for white mummies as fruit bodies of A. apis develops, clean honey bees spread the spores in colony by this behaviour.
- During white mummy stage, fungus continues to develop at hive bottom and when mummies are not removed quickly, spores enter the brood cells are by circulating air.
- The beekeeper can stimulate hygiene behaviour of bees by changing broodrearing conditions, so it is important to adapt the size of hive to strength of bee colony where bees have chance to inspect and clean many brood cells.
- In many cases, method of stimulating hygiene behaviour, as described by European foulbrood control, is sufficient for chalkbrood control.
- The beekeeper should ensure that colony has strong worker population, where hive is well ventilated and free from accumulated moisture.
- At early stages of chalkbrood infection, adding young adult workers and hatching brood, combined with sugar-syrup feeding is helpful.
- The best solution to control is to administer sucrose syrup (1:1) acidification with lemon juice or vinegar or ascorbic acid powder till pH_4, and prevention with application of good management practices in apiary, like selection of appropriate locations, selection of resistant queens and ensuring enough food reserves in hive.

Viral Diseases

Introduction

- Virus diseases of bees are spread throughout the world.
- These diseases undervalued by beekeepers as they can cause serious economic losses if associated with other bee diseases.
- Varroa destructor contributed to increase the incidence of viral diseases and is a passive carrier of bee viruses that are transmitted to bees by mite's saliva.
- The varroa weakens the immune system of bees that allow reactivation of latent viral infections which was present in body of bees.
- Other bee diseases that set the conditions for onset of viral diseases are Nosemosis, European Foulbrood and Amebiasis.

- The transmission of viruses occurs horizontally, but transmission of main bee viruses occurs vertically.
- All viruses are virtually present in apiaries in latent or asymptomatic form.
- Triggering events like other hive diseases or stressful factors, leads to development of infection and death of bees or destruction of colonies and/or affected combs.
- For other viruses, symptoms are general and common such as presence of bees with deformed wings, uneven brood pattern with high mortality and cells with perforated cappings
- The damage caused to colonies by viral infection varies considerably as per number of factors such as type and strain of virus involved, strength of colony, weather conditions, season and food availability.
- Parasitic mites sucking the blood of bees penetrate protection, so increased infestation by parasites is accompanied by increased virus infection.
- Little known viruses like Acute Paralyses Bee Virus and Deformed Wing Virus increases destructive in future.
- As not much is known about life cycle and pathogenity of most virus diseases, there are only a few ways to control them, so reflecting such situation shows only widespread sacbrood.

Types of Bee Virus Diseases

Sac brood

- It is a virus disease where larvae infected with sac brood dies in their sealed cells.
- They become light yellow in colour, with tough skins.
- In this, skin darkens and outer layer becomes loose, forming a sac that encloses watery fluid.
- The brood lies stretched out lengthwise in sealed cell and after death of insect, cell is partly or fully opened, and worker bees remove it from hive.
- The virus is spread in nest by house bees evacuating dead brood and such virus does not survive long, and disease may disappear during honey-flow period.
- Serious outbreaks are not common, and usually no control action is necessary.
- If control is needed, then colony must be requeened.
- Sacbrood disease is perhaps the most common viral disease of honey bees.

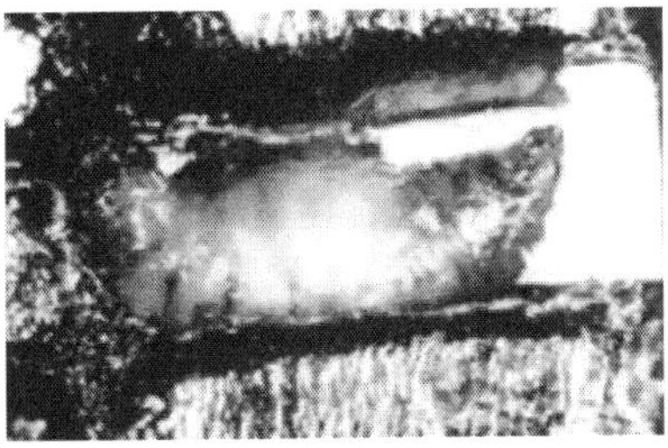

- Such disease affects the common honey bee Apis mellifera and sacbrood disease of Asian hive bee A. cerana.
- A new type of sacbrood virus has recently been reported in Asian colonies of A. cerana. which is highly probable that virus is native to continent and has been with Asian hive bees over longer period of evolution.

Symptoms

- Field inspection to determine whether the pathogenic virus has infected a colony can be easily carried out following symptomology:
- Diseased larvae fail to pupate after four days and remained stretched out on backs in their cells.
- The anterior section of larva, with head and thorax, is first part of its body that change colour from white to pale yellow and finally to dark brown and black.
- On removing larvae from cell, inspector easily observe that their skin is quite tough and that its contents are watery where infected larva has appearance of small, watery sac.
- Dead larvae remains in their cells that eventually dries out to flat scales which adhere loosely to cell floor.

Control

- No chemotherapeutic agent is effective in preventing or controlling sacbrood disease.
- Colonies recover from infection without beekeeper's intervention, if infection is not new to geographic area which mainly depends on hygiene behaviour of bees, that is stimulated as with other brood diseases.
- As disease usually occurs when colony is under shortage of food, food-storage space, unfavorable climatic conditions like dampness during rainy or cold season, unhygienic hive interior, poor queen, infestation with other diseases, etc.
- Beekeeper should deal with severe cases by requeening colony, removing infected brood combs and taking other management measures to restore colony strength like providing food and adding worker population.
- If there is extremely strong infestation, it may be convenient to apply artificial swarm method as for American foulbrood.

Black Queen Cell Virus (BQCV)

- It is a virus which only affects cells of queen bees and is one of the most frequent causes of mortality among the queen larvae.

- The name of virus appears from blackish colour of larval formed and cell walls.
- As worker bee and drone brood can be infected by such disease, they do not develop any kind of symptoms.
- The infection is more common when colonies are affected by nosemosis as lesions of small intestine facilitate the passage of virus in hemolymph.

Protozoan Disease

Nosema disease

- It is among the most serious of bee diseases which is caused by microscopic organism, a protozoan called Nosema apis.
- The disease is one of the most destructive diseases of adult bees, affecting workers, queens and drones alike.
- Seriously affected worker bees are unable to fly and crawl about at hive entrance or stand trembling on top of frames.
- The bees appear to age physiologically as their life-span is shortened and their hypopharyngeal glands ends to deteriorate.
- In this, parasite passes its active reproductive life cycle in digestive cell that is lining in mid-gut of adult bee.
- After entering a cell, parasite multiplies quickly, competing with its host bee for its food supply till reproduction stops after few days, with formation of large number of spores.
- The cell that ruptures and spores enter the bee's through digestive system and passes in bee's droppings.
- The parasite, which is in its resting stage is picked up again by other bee and swallowed.
- In this, spores are germinate on passing from honey sac of bee to mid-gut.
- In this, parasites emerge from spores, pass through the lining of mid-gut and start another phase of intracellular growth and multiplication.
- The spores may remain viable for several months, as long as they remain in the brood combs in the hive.

Cause

- The disease is caused by the protozoan Nosema apis, whose 5 to 7 mm spores infest the bees, are absorbed with the food and germinate in the midgut.
- After penetration into the gut wall the cells multiply forming new spores that infect new gut cells or can be defecated.
- The nutrition of the bees is impaired, particularly protein metabolism.

Symptoms

After the contact of bees with N. apis the following infection symptoms will appear:

- intestinal disorders, such as diarrhea;

- ❖ the bees become unable to secrete the royal jelly;
- ❖ the foraging bees reduce their activity, until it stops completely;
- ❖ in the rare cases in which the queen is sick, egg-laying greatly decreases.

In this, initially there is a slow depopulation, where work decreases while the state of restlessness of colony increases. With this, many bees are no longer able to fly, but can walk with their wings spread out in "K" form, paralyzed, while other bees gather in small groups. The presence of dead bees on bottom of hive with swollen abdomen and legs retracted are found below the chest. The running board of hive entrance and honeycombs will be smeared with diarrhea.

Transmission

The transmission of such bee disease takes place mainly by fecal-oral route by spreading N. apis by droppings sick bees, especially in hive. The spread from hive to hive and apiary to apiary results through:

- ❖ drifting of infected workers;
- ❖ drone displacement;
- ❖ looting of infected colonies;
- ❖ interchanging infected honeycombs from a hive to another;
- ❖ feeding of bees with contaminated honey;
- ❖ use of infected materials or equipment.

Such disease is influenced by many climatic factors such as:

- ❖ bad weather increase chances of infection among bees of same hive as it forces bees indoors;
- ❖ seasonal pattern affects the spread of infection, so during long, cold winters and cold, rainy springs, the bees may not find nectar and pollen;
- ❖ frequent hive inspections with adverse weather conditions can trigger the onset of disease and its propagation as a result of induced stress;
- ❖ presence of other diseases exacerbates the symptoms of N. apis.

Control

- ❖ The disease can best be controlled by keeping colonies as strong as possible and removing possible causes of stress.
- ❖ Colonies and apiaries should receive adequate ventilation and protection from cold and from humidity.

- The bees should have possibility of foraging regularly to defecate in preventing spreading of spores in colony.
- Beekeepers should ensure that their colonies and queens come from disease-free stock.
- Hive equipment suspected of being contaminated by Nosema apis spores should be thoroughly decontaminated by heat and fumigation.
- The best prevention is to change the combs once every two years.
- During normal wax processing, disease spores are killed and only effective chemotherapeutic method for treating the disease feeds the colony with fumagillin at the time when colony encounters stress conditions, during long winter or rainy season.

Precautions

- prevent the infection by using good management practices and special care while selecting apiary location and correct orientation of hives;
- adopting correct wintering measures;
- placing pollen plants near hives which provides protein food to colony in late summer and autumn;
- keeping hives warm during wintering until late spring in cold climates;
- using required number of honeycombs in relation to colony population;
- disturbing bees as less as possible during winter.

Treatment:

- Nosema can be best treated by giving the colony a new set of combs and requeening the hive.
- The affected colony also be given fumigillin with required dosages.
- If medication given cannot be obtained, then option left is to burn the colony to avoid spreading of disease to other hives or even to other apiaries.

4.2.3 Measures against spreading bee diseases

- The beekeeper and honeybee are main agents that spread diseases among bees and between colonies and apiaries.
- Dead larvae, spores and dried scales transported for removal by worker are dragged along the combs before disposed overboard.
- The beekeeper removes combs from weaker colony and exchange them with combs from a stronger colony.
- Sick and weak colonies are united which transfer bees and combs from one apiary to another, thereby spreading diseases.
- The honey contaminated with spores and parasites is fed to healthy colony, or can be dropped by beekeeper where they will be robbed by bees.

- Drones and workers straying in other colonies are guilty of spreading diseases.

For outbreak of diseases, beekeeper should take into consideration following points:

- The apiary must be kept clean as honeycombs, wax, propolis and other hive products should not be thrown away near the apiary.
- The beekeeper must not transfer infected combs from hive to hive or from apiary to apiary, as combs should be exchanged with great care.
- Old hive parts and used apiary equipment bought or acquired from doubtful sources, should be disinfected.
- Unknown swarms should never be accepted when there is outbreak of bee disease, so it is the responsibility of beekeeper to set quarantine apiary at around 4 kilometres from nearest apiary, and to make sure that swarm is disease-free before transporting it to apiary.
- Bees should never be fed with honey from doubtful source.
- If colony dies due to unknown causes, hive should be closed pending an examination of sample comb and the remaining stores in hive should be protected from robber bees.
- Robbing must be prevented, so beekeeper should keep the syrup or food for colony in the hive or in properly designed to feeder to prevent robbing.
- Brood combs should be regularly inspected for signs of disease.
- Hives should be spaced reasonably far apart, so beekeeper should try to arrange the hives so that it will be easy for every bee in apiary to find its way in its own colony, which will help them in minimizing drifting.

Chapter 5

Harvest, Process and Market the Produce

5.1 Study factors during harvesting/extraction of honey bee products i.e., honey, royal jelly, propolis, bee wax, bee venom and pollen.

5.1.1 Introduction

When the honeycomb in hive gets filled with honey and capped with beeswax, then honey is ready for harvesting. Beekeepers should regularly inspect the hives in order to see when honeycomb can be removed. Honeycomb can be removed from hive and is taken to mobile extracting van or central extracting plant. The wax cappings are removed with steam heated knife or special revolving blade before honeycomb is placed in extractor. The honeycomb is placed in revolving baskets where the spinning movement throws out the honey by centrifugal force.

Little or no damage is done to delicate honeycomb with such process and when returned to hive, bees gets immediately set to remove any left-over honey by repairing and polishing each cell in readiness for a new load of honey. Honey collected from extractor is strained and left to stand till air bubbles rise. Bubbles and left-over wax particles are skimmed from surface and honey is ready for bottling. Many beekeepers send their honey in bulk to packing houses while some have their own bottling equipment and sell the honey to retail and wholesale stores.

5.1.2 Factors

The honey needs to be harvested before the bees can consume it for further colony development, but sufficient quantities have to be left to provide for basic needs of the colony. The method consists of collecting honey from wild swarms as no attention was given to survival of robbed colony. Combs with honey, brood and pollen are either consumed directly, without any transformation, or used in production of fermented drinks, so honey from such kind of harvesting is mixed with pollen and brood juice and all other parts of hive.

Colony management

- The exploitation of honeybees is aimed at harvest of honey.
- It is the most rudimentary and ancient method which is still used in some parts of the world, which consists of collecting honey from wild swarms.

- In this, combs with honey, brood and pollen are consumed directly, without any transformation, or used in production of fermented drinks.
- Honey from such harvesting is mixed with pollen and brood juice and all other parts of hive.
- While nutritious, it is not a product that can be included in processing of value added products, other than production of locally appreciated fermented drinks.
- In this, bees are kept in traditional hives made of any kind of suitable, locally available material like tree trunks, rock caves, bark, straw or other plant materials, mud, dung, clay, cut timber or even special cavities in stone or mud walls.
- The harvest time is when the colony has stored the maximum amount of honey.
- In this, different degrees of care takes place for survival of colony during harvesting which depends on type and abundance of bees and knowledge of beekeeper.
- It also uses technique involving dividing of colonies or moving hives as per nectar flows for production to be reliable, involving less expense, but nevertheless remains relatively low in volume.
- Honey produced from such beekeeping can be of good quality as per the knowledge and care taken by beekeeper.
- Product quality ranges from that of most negligent honey robber to that of quality conscious, top bar hive producer.
- Honey extracted by melting combs to separate wax from honey is not recommended as overheating and mixture with old combs spoils the quality of honey.
- Pressing and dripping produces good quality honeys, but even with good comb selection they still contain large amounts of pollen.
- High productivity compared to well managed topbar hives only results from reusability of combs and possibility of migratory beekeeping due to better comb stability.
- Centrifugal extraction allows quick processing of large quantities and produces honey with the least amount of contamination by other hive materials.
- The handling of large quantities allows other processing technologies which foster the production of uniform product with high control of quality standards.

Unifloral honeys

- It shows a sizeable and well-paid portion of European honey market as their production depends on management by site selection and selective harvesting.
- Increasing consumer knowledge and appreciation of honey develops particular market niche for honey that is identified by characteristic colour and flavour, and originating from one or few sources of flowers.

- Differential pricing makes the production from rarer floral sources very attractive.
- In some developing countries, honeys from certain areas are preferred, as not always directly for reasons of floral origin, but sometimes for quality, liquidity, colour or because it looks and tastes the way the most commonly available honey tastes.
- The techniques to produce unifloral honeys are based on possibility of separating honey of one floral period from earlier and later nectar flows on an economically interesting scale.
- The commonly used technique is based on migratory beekeeping where timing the relocation of apiaries and placing and removing of supers is of greatest importance.
- Care also needs to be taken that honey already present in the colony cannot contaminate the colour or flavour of the unifloral harvest.
- Even if production of unifloral honeys is not possible or economically feasible, organoleptic characteristics of honey are still the elements that more than anything else contribute to its consumer appeal.
- It is a good practice whenever possible to avoid harvests that are not much appreciated, i.e. move bees to other areas or leave bitter or otherwise unfavourable honeys to the bees and harvest only at other times of the year.

Contamination during production

- The location of colonies in industrial zones or other areas with considerable air pollution can lead to considerable contamination of various hive products with noxious or toxic chemicals. In many countries, honeybees are used to monitor environmental pollution, as accumulations of certain metals and other substances can be measured in hive products, in pollen and also in honey.
- Agricultural use of toxic chemicals is a source of contamination, so certain pesticides are found in contaminated honey.
- Radioactive contamination in nectars and honeys due to plant uptake of radioactive elements replace normally occurring minerals, where overall content remain relatively low.
- The contamination result from dirty water sources and non-floral sugar sources.
- The worldwide exchange and shipping of honeybee colonies and queens has led to the introduction of new honeybee diseases in many parts of the world.
- Unadapted bees cannot resist the new infections and help from the beekeeper is required. Such help usually involves chemical treatments.

- ❖ If unsafe chemicals are used or even if relatively safe chemicals are applied in exaggerated quantities or at inappropriate times, honey is contaminated.
- ❖ The other source of contamination is treatment of combs against wax moth during storage. All available chemical treatments leave residues in the wax and only abundant aeration (ventilation) for at least a couple of weeks can reduce the hazard, so well ventilated storage without chemicals is preferred.

Contamination during harvesting

- ❖ Many harvesting methods separate bees from their honey where combs can be taken out one at a time and bees may be removed by shaking and brushing.
- ❖ Whole supers can be cleared of bees with a strong air blower.
- ❖ An inner cover or special board with one-way bee escape can be placed below the honey super.
- ❖ Up to one deep or two shallow supers can be cleared in 24 hours, if enough space is available.
- ❖ The method cannot be recommended if colonies are sitting unprotected in the sun, which might melt the combs in the now unventilated supers.
- ❖ The use of unpleasant smelling chemicals to drive bees away is a technique preferred by many beekeepers as it is quick and easy.
- ❖ Some chemicals are illegal for use in many countries, as they leave unpleasant flavours and odours, which are toxic and are absorbed by wax and honey.
- ❖ Excessive use of smoke during harvesting will flavour the honey quickly, no matter which smoker fuel has been selected.
- ❖ In such method, microscopic contamination with soot can be detected.
- ❖ The heavy use of smoke can be reduced by selecting more favourable harvesting times and shorter and more frequent harvests.

Cleanliness

- ❖ Honey in supers of frame hive beekeeping or in broken combs from topbar or traditional fixed comb beekeeping, needs to be regarded as food product.
- ❖ The mature honey is not very stable product, which is altered by, nor, does it permit the multiplication of bacterial or fungal organisms.

- The matured honey can nevertheless be contaminated by either non-biological substances or by potential human pathogens as every caution and care in hygiene should therefore be taken to prevent any form of contamination.
- The general requirement must be taken into account during all processing phases where in the comb, many beekeepers beliefs that honey is exposed to danger of contamination, as the surface area of contact with environment is very large.
- Contact with humid air, with soil, unprotected transportation on dirt roads or in dirty buckets without lid during comb harvesting and exposure to insects and other animals, can adversely affect honey quality.

Processing

- Uncapping is initial step of honey processing which involves removal of thin wax layer that seals the honey cells.
- The wax caps can be sliced off with sharp, thin, long knife or special knives heated by steam or electricity.
- Large numbers of frames are more rapidly processed with partially or completely automated uncapping machines which cut or chop the wax caps with blades, chains or wires.
- In comb harvesting, equivalent step is comb selection, removal of bees etc. and subsequent thorough mashing of combs.
- Processing works further by either letting the wax and honey mixture separated by dripping through screen or by pressing in special honey presses using modified centrifugal extractors.
- The honey frame processing proceeds, after uncapping to centrifugal extraction.
- Extractors range in size from manual to frame model to motorized units extracting more than 12 deep supers at a time.
- More commonly, 24 to 72-frame radial extractors are used for commercial enterprises.
- The smaller units for part-time beekeepers can be made from recycled materials.
- Though honey can be extracted faster and more completely at higher temperatures, combs will become softer and might break, so extraction temperatures should not exceed 30^{0}C.

Purification

- The next step in honey processing is removal of any impurities like wax particles, debris and air bubbles during extraction.
- There are two practical techniques used, which can be settling and straining.
- In settling, honey is left in large container, so that impurities can separate as per their specific weight, to the surface while mineral and metallic particles drop to the bottom.
- The surface scum can be removed carefully, or honey can be drawn off near the bottom for bottling without disturbing either surface scum or bottom sediment.

- Settling velocity varies with particle size, container size and honey viscosity, i.e. moisture content and temperature.
- At temperatures of 25-30°C settling is quick and can be completed in few days where tanks needs to be well covered to avoid excessive contact with air.
- The process can be accelerated by allowing honey to flow through special buffer tanks prior to filling into settling tanks.
- In buffer tanks, honey is heated by water jacket, similar to water bath which is forced to flow up and down through several compartments of which impurities remain at surface
- Subsequent settling frees honey of air and foam and if containers are big enough, it allows some mixing of extractions from various colonies to have certain degree of uniformity of end-product.
- The disadvantage is cost of containers for extra storage lasting several days, which in large operations that needs many large tanks and large amounts of extra space.
- Straining can be used instead of or in addition to settling as it is more frequently used in larger processing plants, where many tones of honey are processed every day and where it is inconvenient and uneconomic to immobilize honey for as long as is required for settling.
- Strainers are metallic screens, which are covered with fine nylon mesh or nylon sack filter submerged in tall, narrow tank.
- In this, sack-like filter can be made of many layers of increasingly finer metal screens.
- Such filters have advantage of large filter surface which can be submerged to avoid any further inclusion of air.
- The finest mesh size used has holes of 0.1 - 0.2 mm diameter where temperature is around 30°C.
- Finer filtering is done in association with pasteurization and heating of honey to 77 -78°C which removes fine materials, including pollen, to delay crystallization for long as possible.
- Such filtration needs high pressure filters with diatomaceous earth as it requires heating to remove some natural ingredients such as pollen.
- Such honey cannot be sold as table grade honey in EEC countries as consumers in such countries regard it as inferior in quality, while it is preferred quality for supermarkets and other large marketing chains which requires product with longer shelf-life in homogeneous liquid state.

Moisture content

- Moisture content of honey is important quality parameter as it affects storage life and processing characteristics
- Even though moisture can be removed after extraction, only completely ripe honey should be harvested, i.e. combs with more than 75 % of the honey cells sealed.

- For having 75 % of honey cells sealed before very end of nectar flow, the colony has to have sufficient super space for storing incoming and ripening honey.
- When average atmospheric humidity is not much above 60%, moisture content below 18% will be expected in the honey.
- In more humid climates, sealed cells contain honey with more than 24% even 28% moisture content.
- Combs with fresh nectar should never be harvested, as it can dilute and spoil the whole harvest unless of course the purpose of harvesting the honey is making beer.
- Post-harvest reduction of moisture content can be achieved by leaving honey supers in warm rooms at 30 to 35 ^{0}C and circulating warm air through them.
- During this period, surface area of comb relative to honey mass is large and does not need any extra equipment for efficient evaporation.
- In relatively cool climates, circulation of air heated to 35 ^{0}C reduces moisture content in open honey cells by 1% to 3 %.
- The relative humidity of air at 35^{0} C has to be controlled, but if it is more than 60%, then aerial moisture needs to be removed by dehumidifier.
- In tropical climates, air temperature will be higher or prior dehumidification of air will be necessary which needs small, specially sealed room and dehumidifier.
- Post-extraction moisture removal is more involved, but small scale methods are available.
- A cheap small scale honey drier, for solar heating, where hot air is conducted over thin film of honey that runs on inclined surface.

Prevention of fermentation

- Fermentation is microbiological alteration to which honey is susceptible.
- Only osmophilic yeasts can grow in high sugar concentrations, but their presence is ubiquitous in honey, nectar, hive interiors, dust and soils.
- The rate of multiplication increases proportionally with increasing water content, up to a certain point.
- Below 18 % moisture content, there is little probability of fermentation, but at concentrations below 17.1 %, the risk of fermentation cannot be completely excluded, so such aspect of fermentation depends on factors like quantity of yeast and other growing factors as honey temperature and distribution and availability of water following crystallization.
- Appropriate and expensive cold storage, careful production techniques, can prevent fermentation.
- If, after all precautions and care, honey cannot be harvested at less than 18% water content, then in such case, excessive moisture should be removed.
- The methods, if carefully used and if honey has not yet fermented, will able to prevent fermentation without degrading the honey.

- ❖ Method based on pasteurization and destruction of yeasts where osmophilic yeasts in honey dies only after few minutes of exposure to temperatures between 60 to 65 ^{0}C.
- ❖ If honey is heated and cooled quickly, with special heat exchangers feasible only on an industrial scale, very less damage occurs to honey.
- ❖ The pasteurization treatments prevents fermentation and postponement of crystallization.
- ❖ Small quantities of honeys with high moisture content do not justify complicated and costly pasteurization, or drying, but should be directed towards market with immediate consumption, for processing in other food items or fermented drinks.
- ❖ The honey obtained should not be considered for shipping over long distances as containers might explode.
- ❖ Careful heating in water bath to wax melting temperatures (about 65 ^{0}C) and subsequent cooling in water bath with running water prolong honey storage life.
- ❖ For small quantities, it is acceptable between spoilage by fermentation and loss of quality by heating.
- ❖ Under most circumstances, water baths are overheated and honeys are not properly stirred and cooled down rapidly enough.
- ❖ Pasteurization on small scale is recommended for emergencies and not as a routine procedure as it is used in many places.
- ❖ The pasteurized honey needs to be bottled hot in clean environment to prevent re-infection with omnipresent yeasts.

5.1.4 Process of Bee Harvesting

Following are the steps which beekeeper needs to follow during bee harvesting which includes:

Step 1: Removing Honey Frames from Hives

- ❖ The initial step in bee harvesting is to take out the frames of honey from the hives.
- ❖ The number of frames depends on many factors including strength of the colony, weather, and amount of nectar available to the bees.
- ❖ Beekeeper is advised to wear protective gear while removing frames of honey from hives.

- The boxes filled with honey can be heavy so beekeeper should remove them one frame at a time.
- After removing the frame, beekeeper should gently brush off the bees from the comb and place them in a cart that is wheeled to garage.

Step 2: Using Equipment

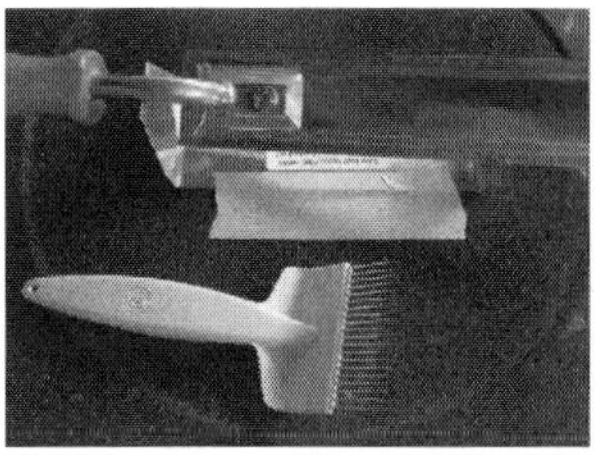

- Beekeeper should used certain harvesting tools to extract more frames at one time.
- Beekeeper should use heated knife and capping scratcher for remove wax cappings on the frames.
- A tarp is taped to the floor to keep things clean, so beekeeper should use bigger bucket with strainer for extracted honey.

Step 3: Uncapping the Frames

- Honeybees preserve the honey by capping it in wax cells.
- To extract honey, beekeeper should remove top of the cells or caps.
- Many frames have honey on both sides, so beekeeper should uncapped each side oof the frames.
- Beekeeper should use cap scratcher to pick out any cells that have not been opened.

Step 4: Spinning the Frames

- In this, four uncapped frames are kept in extractor and are held in a metal mesh basket.
- The lid is closed and handle cranked for a minute or more.
- In this, honey is forced out of comb and drips down the inside of extractor.
- Once spinning stops, frames are taken out and flipped so that honey is removed from other side of frame.

Step 5: Flowing of Honey

- When honey starts filling up the bottom of extractor, beekeeper should open up the valve and allow honey to flow in waiting bucket.
- Now beekeeper need to quality test the extracted honey.

Step 6: Filter and Jar

- With much uncapping, spinning and tasting, the bucket gets filled with honey.
- The honey gets filtered by paint strainer.
- The bucket has latch near the bottom that can be opened and closed for filling jar

Step 7: Melting the Wax

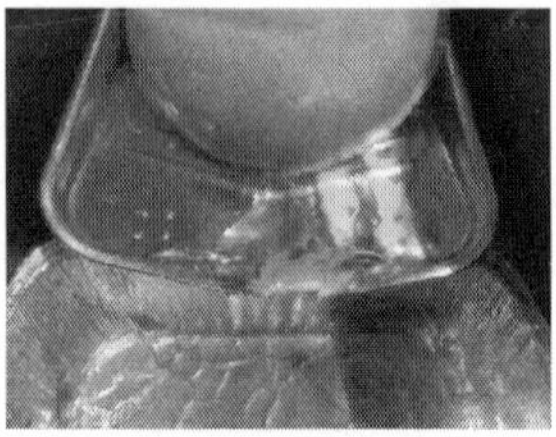

❖ There is lot of honey and wax in cappings that were cut off the frames.

❖ By using wax melter, the wax gets melted which is used for making lotions and furniture polish.

Step 8: Keeping Honeybees

❖ Beekeeper should keep extracted honey in safe place in safe jars, so that it can be used for eating, baking and making lots of other things.

5.1.5 Bee Products

❖ Bees with their pollination activities maintain biological balance in nature and allow many animal and plant species to thrive by providing bee products that are entirely natural food source.

❖ Honey, bees wax, royal jelly, bee venom, propolis and pollen are the important bee products.

Beeswax

❖ It is the material which is used by bees to build their nests.

❖ It is produced by young honeybees that secrete it as a liquid from special wax glands.

❖ Worker bees secrete wax when they are around 14-18 days old.

❖ When wax gets in contact with air, it gets hardened and forms scales, which appear as small flakes of wax on underside of bee.

❖ Bees use wax to build hexagonal cells which makes their comb, that serves as strong and efficient structure.

❖ Bees use comb cells to store honey and pollen as queen lays her eggs in them while young bees develops in it.

❖ Beeswax is produced by all species of honeybees, but wax produced by other species have slightly different chemical and physical properties.

❖ It contains 70% alcohol's and 75% fatty acids with 13%-15% of free acids, 12%-15% saturated hydrocarbons and vitamin A having specific gravity 0.95 and melting point 65°C.

❖ Beewax is obtained from cappings, which gets collected during honey extraction.

❖ Wax is obtained from old combs that are unfit for use and from combs damaged during honey extraction.

- ❖ Best grade wax is obtained from cappings where recovery per cent is higher.
- ❖ It is mainly used by candle industry and are used for preparing comb foundation sheets.
- ❖ It is used by cosmetic, pharmaceutical and perfume industries.
- ❖ It is used in manufacturing of cosmetics items, polishes for boots, furniture, lubricants, paint and varnishes, inks, electrical insulating apparatus and candles.

Propolis

- ❖ It is gathered by bees from resinous exudes of tree.
- ❖ In bee colony, propolis is used for sticking frames, sealing cracks and crevices as it is contaminant of comb wax.
- ❖ It is obtained by scrapping it from frames and contains 55% resins and balsams, 10% ethanol and scented oils while 5% pollen.
- ❖ It is used in preparing ointments for treating cuts, wounds and abscesses in cattle.
- ❖ It is mixed with vaseline to smoothen burns.

Royal jelly

- ❖ It is secreted by gland of nurse bees that are 6-12 days old when glands are fully active.
- ❖ It is very nutritious food and is fed to young worker larvae and queen larva and adult.
- ❖ The jelly is milky or light pale in colour.
- ❖ It contains 18% proteins, 6% lipids, 17% carbohydrates, 1% ash along with vitamin A, B and C, iron, copper, phosphorus, silicon and sulphur.
- ❖ The queen cell is trimmed to level of royal jelly which after 2-3 days of grafting, larvae gets gently removed with forceps and royal jelly is removed with royal jelly spoon which gets stored in refrigerated conditions.
- ❖ In Apis mellifera, 200mg of royal jelly is obtained from queen cell.
- ❖ It is responsible for queen determination and is nutritious food for humans.

Pollen

- ❖ It is collected by pollen trap from ingoing pollen foragers.
- ❖ It is rich protein source for human diet.
- ❖ It provides humans with valuable materials such as honey and beeswax.
- ❖ The honey bees are useful in aiding pollination, so value of bees in pollination of crops is ten to twenty times the value of honey and wax produce.
- ❖ Certain crops like apples, alfalfa and clover almost entirely depend upon bees for their pollination.
- ❖ Even among some regularly self-pollinated crops, yield is considerably increased after visit of bees.

5.2 Study tools/equipments required for harvesting/extraction of honey bee products

Honey Harvest Tools

1) Beekeepers suite and accessories

- ❖ It involves mesh helmet and folding veil, with some layers of clothes.
- ❖ No specific design of protective clothing for beekeepers exists, but white overalls are occasionally worn.
- ❖ Many clothing suitable for field work is good for beekeeping, as light-coloured, smooth materials such as cotton are preferable as they are cooler and create less risk of antagonizing the bees.
- ❖ The beekeepers feel more secure in handling bees if they are wearing gloves as bee gloves are made of tightly-knit cloth and/or soft leather which can cover the forearms upto elbows.

2) Smoker:

- ❖ The bee smoker is used to calm the bees which consists of metal fire-pot with funnel-shaped cover and bellows.
- ❖ The smoker is available in many models and is equipped with shield for protection against the heat generated.
- ❖ A smoke-releasing fuel is burned in fire-pot, and air is injected in the pot by operating bellows where the smoke gets directed at bees by funnel.

3) Frame super

In this, frames with honey combs is kept for transportation

4) Sting resistant gloves

5) Hive tool

- ❖ Honeybees use propolis to seal frames and covers to hive bodies and supers.
- ❖ To separate various pieces, beekeeper prizes them apart with hive tool, which is useful in scraping excess propolis or wax from hive parts.

- ❖ It is convenient equipment which beekeepers use while working with their hives.
- ❖ It consists of length of iron or steel, flattened at one end.
- ❖ It is a good commercial hive tool which is made of spring steel.
- ❖ It is used to move the frames, scrape wax, etc.

Honey Extraction Tools

Introduction:

- ❖ Honey extraction is the main process in beekeeping which involves removing of honey from honeycomb so that it is isolated in liquid form.
- ❖ Honey is stored by honey bees in beeswax honeycomb which is framed in bee hives in wooden structure known as frame.
- ❖ Honey frames are harvested in late summer, when they are filled with honey where cells are capped over by bees for storage.
- ❖ Each cell with honey is sealed with capping which is made of beeswax.
- ❖ The extraction of honey is done in special room or honey house that is heated with necessary tools and can be washable.
- ❖ The room must be sealed, as bees will eagerly try to enter and gather the honey.

Tools:

- ❖ Beekeeper uses certain tools during the extraction process of honey which includes:

1) Heated knife - to unseal honey cells

2) Uncapping knife

- ❖ Uncapping knives are used for slicing quickly and cleanly through wax cappings on the honeycomb.
- ❖ These cappings create an airtight seal on cells that house the honey, making it hard to extract.
- ❖ With electrically heated uncapping knife, beekeeper can remove the lids so as to extract the honey from the combs.

3)Tub for wax/honey

4) Extractor

- It is a cylindrical piece of equipment which is used to extract honey.
- It is a device that uses centrifugal force in squeezing honey out of comb.
- It comes in various sizes and styles and are designed to meet virtually any need and budget.

5) Food-grade bucket - to catch honey out of the extractor

6) Honey strainer

- The extracted honey requires to be strained before bottling, so strainer removes the little bits of wax, wood and sticky bee.
- It is usually made of stainless-steel.

7) Containers

- As honey is delicate and perishable product, so after it has been extracted and strained, it must be properly stored.
- As honey can absorb moisture from air and can reacts with metals or can changes colour if exposed to light over long periods, and can also result into change of taste on heating, so it must be neither heated nor stored in excessive warm room.
- If honey is not bottled for sale immediately after extraction, then it should be stored in clean, non-corroding, tightly-sealed containers strong enough to withstand its weight.
- Glass-lined steel drums, off-white opaque heavy-duty polyethylene cans and lacquer-coated or galvanized-iron cans are among the popular containers used.

5.3 Proper methods of harvesting/extraction of honey and other honey bee products

5.3.1 Time of Harvesting

There are many factors that goes into decision regarding the time to harvest honey from hives.

- The beekeeper should give bees the required time to collect the nectar and process it into honey by lowering the moisture content of nectar to be below 18% for fast process and harvest before it gets ripened by bees.
- If high quality honey stores well or is for sale, then low moisture is required, so beekeeper should not extract honey till there are at least 75% of cells in honey combs which are capped. By extracting one uncapped honey frame after three capped frames

of honey, overall moisture content tends to be below the level where fermentation takes place.

- ❖ The beekeeper should wait till all blossoming plants have finished blooming so as to maximize honey crop by harvesting late in season as cold temperatures make honey in combs thick and makes it difficult to extract.
- ❖ Robbing pressure is also very strong in cold, so with no other sources of nectar to forage on and in winter right around the corner, honey supers should be covered and protected from marauding bees as they can take off the hives.
- ❖ The extra effort involved in defending crop from robbing bees reduces the interest of honey harvest.
- ❖ It is not usually preferable to put off the honey harvest till late in season as some of treatments used for pests and diseases are temperature sensitive, so waiting too long till average daytime temperatures will not be favorable for effective treatments.
- ❖ When colony has stored far more honey than they can reasonably expect to consume during long winter, its best to harvest the excess rather than leave it on hive for winter as certain beekeepers wait to harvest the honey in spring rather than in autumn to make sure that bees have enough honey for winter.
- ❖ In sunny day between 9:00 AM to 4:00 PM hours, harvesting of honey during the time frame is good as there are naturally less bees to deal with.
- ❖ The time of season in which beekeeper harvest can also make big difference in the yield and quality of honey as in late summer and early fall, bees stop producing honey to feed the queen where more clusters are left empty, so beekeeper should harvest the honey earlier in the season.
- ❖ The harvesting of honey should be done two to three weeks after prime nectar flow occurrence when hive is at its heaviest.

5.3.2 Method of harvesting

- ❖ Honeys produced in different places and at different times of the year can be quite variable, offering a range of flavours.

Using Extractor:

The honey extractor is a device which extracts honey without damaging bee combs. It contains a cylindrical drum holding a frame basket, and by spinning it around honey is extracted by centrifugal force. It is a great method as in this, wax and comb remain intact which can be returned to hive.

Tools required by beekeeper for extracting honey using honey extractor which includes:

- ❖ An extractor
- ❖ A hot knife for cutting off wax cappings on honeycomb
- ❖ A strainer with fine pores
- ❖ A bucket for collecting extracted honey
- ❖ A measuring cup for pouring the extracted honey in small containers
- ❖ A cookie sheet for catching wax cappings that are cut off with hot knife
- ❖ Honey jars for storing honey

Extraction process:

- ❖ It is one of the easiest and fastest ways to take honey out of cells, so beekeeper requires some special and expensive equipment to use such method.
- ❖ Before beekeeper put the frames in extractor, he/she needs to remove the wax caps from the frames to allow the honey to come out.
- ❖ Beekeeper can use special uncapping knife for this or can use 2 bread knives with one knife sitting in hot water while using other and When one knife are used which has cooled, he/she has to switch between the knives.
- ❖ The beekeeper has to make sure that he/she has to dry the knife which is kept in hot water with towel or calico before using it.
- ❖ The beekeeper should not put water in the honey, but should keep the cooled knife in hot water for knife to be heated up.
- ❖ To uncap the wax seals or caps, the beekeeper has to put a nail in board so that pointed end sticks up through the board.
- ❖ The beekeeper should keep the board over a large basin and rest one end bar of the frame on pointed end of the nail and hold the top bar of frame towards himself and should start from the top.

- ❖ The beekeeper should use steady gentle sawing motion while cutting downwards and should fill frames of honey.
- ❖ The beekeeper should see that the more wax he/she leaves on the comb, less wax the bees will make while giving up empty frames back to them.
- ❖ The basin will catch the wax caps and the honey when beekeeper cut the caps off the frame.

- The wax caps will still have plenty of honey in them that beekeeper can squeeze out later and should turn the frame around and cut the wax caps off of the other side.
- When wax caps are off both sides of frame, then beekeeper should put it in wire baskets of extractor while all wire baskets in extractor results in uncap frames and should spin the frames by hand crank on extractor.
- After beekeeper have spun the extractor for about 20-30 seconds, he/she should turn the frames around and spin out the honey on other side.
- Using some extractors, beekeeper should remove the frames and turn them around to spin them again to get honey out of other side.
- When beekeeper have finished spinning the frames, they look little wet and the honey will not be out from the frames.

- With this, there will have some honey in bottom of extractor, so beekeeper needs to open the honey gate or valve on extractor and pour through a strainer in buckets to store which will remove the pieces of wax or bee parts that have left while extracting the honey.

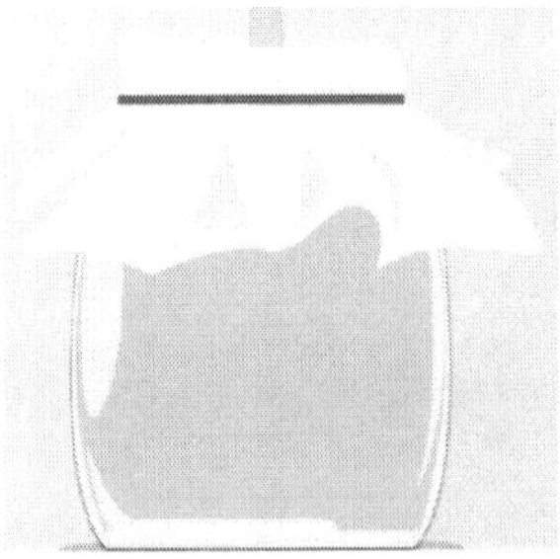

Using Spoon:

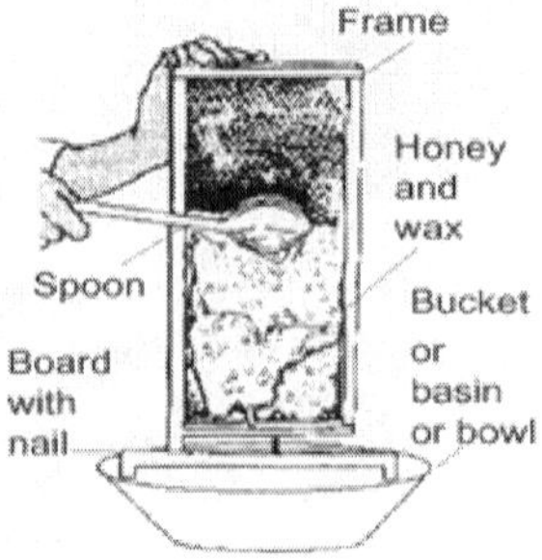

- If beekeeper did not have an extracting machine, he/she can take the honey out of frames using a spoon by holding the frame in same way as cutting wax caps with knife.
- Since beekeeper uses spoon to scrape the wax and honey in basin, so only scrape in as far as wax foundation sheet or beekeeper will make holes in wax will spoil the frame.
- Keep the wax and honey scrapings in strainer and squeeze the honey out of the wax.
- Beekeeper should not need expensive equipment for such method, but he/she can damage the wax frames which are very expensive to replace. The bees must make beeswax to build cells what beekeeper destroyed while spooning out honey.
- The bees should eat a lot of honey to make new wax, so beekeeper won't get as much honey in an year.
- Such method is messy and sticky as the work is hard to get all honey out of wax.

Using Press:

It is another method where wax gets damaged to remove the honey from the comb. For this, beekeeper should:

- Cut the wax from the frame.
- Remove the wires from the wax.
- Wrap the pieces of honeycomb in calico.
- Squeeze the comb in the calico to break the wax.
- Twist or wring the calico to press the honey through cloth by using two people to twist the calico to get honey out of comb.
- It is faster if beekeeper makes a wooden press to squeeze out most of wax.
- The beekeeper should strain the honey and put it in large, covered, airtight container for few days.
- Any extra bits of wax or pollen will float to the top of honey, so beekeeper needs to remove these bits before storing the honey.
- The beekeeper should press the honey which is messy and sticky, as all the honey cannot be removed from the wax.

The biggest disadvantage to spoon and press methods is that beekeeper must break or damage the honeycomb. It takes the bees much time and energy to replace this lost wax.

5.4 Study factors affecting quality of bee products

5.4.1 Introduction

The best known primary products of beekeeping are honey and wax, but pollen, propolis, royal jelly, venom, queens, bees and their larvae are also marketable primary bee products. Most of these products can be consumed as they are produced by the bees.

5.4.2 Bee Products

The main bee products that are used for human consumption and use are:

Honey

- Honey bees suck the nectar from flowers and store it in a stomach-like organ called a honey crop.
- When bee returns to the colony, another bee takes the nectar and spreads it over the wax honey comb to help water evaporate.
- The next bee adds and invertase enzymes help in break down of sugar molecules.
- Once it becomes thick it is sealed in a cell with a wax cap.

Bee Wax

- It is a natural secretion from wax glands on sides of body of honey bees.
- It is used mainly as building block for bees' honeycomb cells where young are raised and honey and pollen are stored.
- For stimulating beeswax production, honey bees feed themselves with honey and huddle together to raise temperature of cluster.
- It is used in many products such as skin care, candles, furniture polish, batik-making,etc.

Pollen

- Pollen grains are small, male reproduction units (gametophytes) formed in the anthers of the higher flowering plants.
- It is the male seed of flower blossom which are collected by honey bees and mixed with bees' digestive enzymes.

- Bee pollen is low in calories but rich in proteins, amino acids, vitamins, minerals, enzymes, fatty acids, carbohydrates, and bioflavonoids which are anti-viral, antibacterial helpful in lowering of cholesterol, stabilising and strengthening capillaries.
- It rejuvenate the body, stimulates organs, enhances vitality and accelerate rate of recovery which makes it popular tonic among athletes and sportsmen.

Propolis

- It is a bee glue which is a mixture of beeswax and resins collected from leaf bugs and twigs.
- It is used to line nest cavities and brood combs, seal cracks and reduce the size of hive entrance.
- It has antibacterial and antifungal properties.

Royal jelly

- It is a protein rich substance that is fed to larvae.
- It is more given to queen larva, which cause her to grow larger than other bees.
- It is made from digested pollen and honey which contains sugars, fats, amino acids, vitamins, minerals and proteins.

Venom

- It is used in bee sting and is made of complex mixture of proteins.
- It gives benefits to humans.

5.5 Processing, packaging and storage of extracted honey

- Honey is at its peak quality when it is properly cured and sealed in comb by honey bee.
- When honey is converted from raw state by humans to particular state, deterioration begins.
- The extent of deterioration depends on processing methods used between the time when honey is extracted from comb and its use by consumer.
- It is responsibility of industry to provide top quality product to consumer if acceptance is to be expected.
- As much honey harvested is extracted from combs in an extracting plant, beekeeper should equip the plant so that the operations can be done in most efficient manner to provide high-quality product for market.

5.5.1 Processing

- Processing honey crop beyond the extraction stage is done by producer/packer.
- Regardless of where these operations take place, producer/packer provide consumer with high-quality product.
- It is important that heating to be controlled as flavor, color and aroma of honey can be seriously impaired by excessive temperature over given period of time.

Different types of Honey Processing

Cut-comb honey

- The simplest processing of such type of honey is to remove honeycomb from frame hives, top-bar hives or traditional hives and sell/consume it as cut-comb honey.
- When producing such type from frame hives, it is necessary to use wax foundation that does not contain strengthening wires and is thinner than that normally used in wired frames.
- The process involves collecting pieces of sealed and undamaged honeycomb, cutting them into uniform sized pieces and packaging them carefully in bags or cartons to avoid damaging the honeycomb.
- As honeycomb is unopened, it is readily seen to be pure, and has finer flavour than honey that is exposed to air or processed further.
- Cut-comb honey have high local demand and fetch higher price than processed honey.
- The honeycomb is easily damaged by handling and transport, which makes distribution for retail sale more difficult, so it needs protection by packaging materials that will absorb shocks or vibration and packs should be carried carefully and not stacked, thrown or dropped to avoid damage to honeycombs.

Strained honey

- This honey is processed to minimal extent and is sold locally.
- It is prepared by removing wax cappings of honeycomb using long sharp knife that has been heated by standing it in warm water.
- The honeycombs are broken into pieces and honey is strained to remove wax and other debris.
- A coarse strainer is used at first to remove large particles, and honey is strained by successive finer strainers like cotton/muslin cloths.
- The clear honey is collected in clean, dry container.
- When most of honey has drained, combs are squeezed in cloth bag to remove much of remaining honey.
- The wax is collected and formed in block by melting it gently in warm waterbath or solar wax extractor.
- Such beeswax byproduct has high value as wax polish or for candle-making.
- The strained honey can either be dispensed from collection pan in customers's own containers or packed in glass jars or plastic bags for sale.

Grading

- Honey grading is mostly done for exporting the product in market.
- Grading is based on characteristics of honey such as moisture content, freedom from foreign matter and flavour.

- Honey is graded on a scale of 1 to 3, with 1 being the highest quality.
- Grocery stores usually sell grade 1, as lower grades are used in commercial food processing.
- Honey is classified as per the colour while prepackaged honey is classified as white, golden, amber or dark.
- Colour of honey is determined by type of flowers visited by bees.
- The light-coloured honey is more mild and sweet, while darker honey has stronger taste.

5.5.2 Packaging

- When honey is extracted from top-bar frames, the frame is kept over a dish and thin layer of wax capping is cut from bottom to top of frame and is allowed to fall in dish below, where frame is turned and capping on other side is removed.
- Honey that is stuck to wax cappings is strained using cloth bags where frame is placed in a honey extractor.
- Honey extractors can be manually or electrically operated, as per the scale of production and can be tangential or radial type machines.
- The frames are turned so that other face of honeycomb faces outwards, and machine spun till this side is empty to prevents the inner part from bursting by empty outer combs and to prevent combs from breaking.
- The honey is collected in pan made from food grade plastic or stainless steel, and filtered through nylon/stainless steel filter unit having fine filters, as honey moves to outside of filter unit.
- The clear honey is collected and packaged in glass or plastic containers and labelled.
- The package should be moisture proof to prevent honey picking up moisture from the air during storage, as customers regard the colour of honey as important quality characteristic, the containers should preferably be transparent so that customers can see the product.
- Glass jars with screw-on lids or plastic pots with heat-sealed foil or plastic lids may be used.
- As per the intended market, honey can be bottled directly in small containers for retail sale or in large drums for storage/export.
- In an effort to appeal to wide range of consumers, honey is packaged in containers of different sizes and styles which include glass jars, plastic tubs and squeezable jugs.

Labelling

- The label on honey container in a retail outlet should include the word "Honey" or possibly, an indication of floral source.
- It should have grade, net weight, name and address of honey dealer and registration number of packer, with nutrition facts table.

- The label identify the country of origin and shows whether honey is creamed, liquid or pasteurized.
- The information marked on the label should be normally in English or in any local language with minimum type size in specific areas of label where it's easily found and read by consumers.
- Honey sold at a farm gate or farmers' market does not need to meet the same labelling requirements because it's usually coming directly from the producer.

5.5.3 Storage of Honey

- It is important to store honey in air tight containers as it helps in protecting water content, as if water is allowed to evaporate, it gets extracted from honey and crystallize quicker.
- If the water is inside the honey, then there are chances of fermentation as honey will not ferment if its water content is under 17.1%.
- For best shelf stability of honey, it should be stored in glass jars as plastic containers allow honey to lose water content or lead to leeching of chemicals.
- For best storage in plastic use HDPE plastic. Stainless steel containers are also approved for long term bulk storage. Avoid all metals that are not stainless steel as corrosion will contaminate your honey.
- Honey in bulk containers, should be stored in a dry place at 70°F as possible.
- Long periods of storage above 70° will damage the honey, same as excessive heating.
- Storage of unheated honey at 50° to 70° is inductive to granulation and fermentation which is true for honey packed in bottles and containers.
- The deterioration of honey during storage can be prevented by maintaining storage temperatures below 50°.
- Honey stored at freezer temperatures, 0° to -10°, for years cannot be distinguished from fresh extracted honey in color, flavor or aroma.

Handling Honey

- Honey is sold wholesale to packers in gallon cans, drums or in bulk.
- For many years, the gallon can was main wholesale container which small beekeepers still uses.
- As single can may be handled by hand, they are more easily handled by pallet truck.
- Larger quantities of honey are more easily handled in big-gallon drums.
- The drums are more durable than cans and are reusable.
- Many industrial hand and power trucks are available for handling drums.
- If drums are to be stacked, motor-driven lift truck is required, so all commercial bottlers of honey are equipped to handle both cans and drums.
- Some beekeepers built large tanks for storage of 1,000 gallons or more of honey, which is pumped in tanks on trucks for transfer to bottlers.

❖ Many companies use air pressure to transfer honey in and out of truck tanks.

5.6 Identify suitable market platform for honey bee products

5.6.1 Introduction

❖ Beekeepers have to focus on production and should concentrate on production of bee products with relative high output for selling more bee products and earn more money.

❖ A production focus is good, but without selling bee products there would be little or no production.

❖ Selling bee products involve transactions, where exchange of honey for money is done with customer/retailer or anyone who needs honey.

❖ The transactions take place in markets, so it is good for beekeeper to understand about market.

❖ Market is where bee products are sold and consumers are able to choose and buy the bee products.

❖ Market is a place where suppliers/beekeepers/retailers meet demanders/consumers.

❖ Other markets are those that occur when a beekeeper sells his/her bee products to processors/retailer where supplier is beekeeper and demander is processor/retailer and demanders are called as business consumer

5.6.2 Supply System for bee products

Supplier → B.keeper → Rural Merchant → Rural Wholesaler → Urban Retailer → Consumer

❖ The supplier sells to beekeeper which then sells to rural merchant, which in turn sells to rural wholesaler and then to urban retailer, who sells the honey/product to consumers.

❖ During transactions at each stage of supply there are exchanges taking place as there are many markets.

❖ Since beekeeper sells his/her honey comb to rural merchant, so beekeeper is supplier and rural merchant is demander, so such transaction serves as business market. If beekeeper sells directly to consumers, through stall in village market, in such case the market would be consumer market. In such case, beekeepers can sell to business markets or to consumer markets or both.

❖ In markets, exchanges take place and bee products are exchanged for money as beekeeper needs money to survive, hence beekeeper has to offer certain quantity of products for making enough money to survive.

❖ Finally what the beekeeper has to earn is profit as profit is difference between what the business had paid to make the bee product and make it available to consumer, costs and price it receives for bee products, as without profit, bee keepers will not generate enough money to survive and continue in operation.

- Profit is the sum of money over and above the costs that the beekeeper has paid to produce and sell the bee product, so beekeeper needs to obtain a price for product that is higher than his/her costs.

5.6.3 Factors for Market Demand

In consumers markets, demand is influenced by:

Tastes:

- All consumers have different tastes, as one consumer likes honey comb, while another may like processed honey in jars.
- Tastes also change, as one consumer want more honey and less sugar, so changes in taste will change demand and price.

Season:

- Consumers change what they want to buy as per the season.
- In festive season, consumer may demand more processed honey to make cakes while in hot season, consumer may demand for less honey.
- The fluctuation increase or decrease in quantities in demand for product will affect the honey price.

Location:

- Consumer who lives in different places will have different demands for products.
- If a consumer living in urban cities prefer processed honey to honey comb, while consumer in rural areas prefer comb honey.
- One village market in a location may have different demands for different types of honey than another village market in another location.
- Consumers in far away markets in other countries have different demands for products which affect quantities demanded and price of bee products..

Income :

- Consumer demands for bee products depend on what consumers earn, or on what wealth they have to determine how much they are able and willing to spend to buy the products.
- If consumers have lot of wealth, they will buy more products, but will also want to buy more expensive of luxury products.
- If consumer buy local honey, but they may also buy expensive honey that comes from other countries, then amount of wealthy consumers will affect quantities demanded and price.

Population:

- The increase or decrease in population will affect consumer demand.
- If area is more populated than others than demand will be higher.
- If a region has many people they will be demanding more bee products which will affect the price.

Age:

- The consumer demands for products will change as a result of their age.
- In areas where there are many young children, demand for honey maybe be very high.
- In areas where there are many old people, the demand for propolis may be high as propolis serves as good medicine, but at the same time is very digestible.
- Age will affect quantities demanded and price.

Price:

- Price of product ultimately affect the demand for the product.
- If the price of processed honey is high, consumers will buy less processed honey.
- If prices of comb honey are low, then consumers may buy more comb honey.
- Prices will affect quantities demanded.

Future prices:

- The consumer demand will change as per the prices expected in future.
- If high prices are expected in future, consumers will buy more quantities of bee products now.
- During honey harvest time, the year prices gets low if honey harvest is good.
- As time passes and honey gets consumed, the prices will again start to rise.

Quality:

- Consumer demand will change in regard to the quality of bee products.
- If the consumers think that quality of honey is not god they will not buy it which affects the price of honey and quantities bought.

Similar use of products :

- Consumer demand for products that are similar in usage will change when price increases as compared to another.
- In case of processed honey and sugar, if the price of honey rises, consumers may prefer to purchase sugar as an alternative.

5.6.4 Searching for Bee Product Market

- Honey is a rich source of antioxidants and flavonoids which reduces oxidative stress and inflammation in body while protecting against heart diseases and cancer.
- It has antibacterial and antiseptic qualities that accelerate wound and burn healing by providing moist environment and increased flow of nutrients to affected area.
- Producing honey for profit depends on successful marketing, so beekeeper should conduct market research which determines competition, amount of honey to sell and type/forms in which the customers prefer their honey.

Markets for honey and bee products are extensive, so beekeeper should plan to start small and expand as market demand, increases and develop better understanding of markets for their products for which well-organized marketing plan should be designed which includes:

- ❖ Production of high-quality honey
- ❖ Attractive containers and labels
- ❖ Effective advertising program
- ❖ Dependable customer service

Most honey across the world gets extracted and sold in form of liquid, but it can be prepared and marketed in following ways:

- ❖ Extracted honey
- ❖ Section-comb honey
- ❖ Cut-comb honey
- ❖ Chunk honey
- ❖ Finely crystallized or creamed honey

Once beekeeper/owner is ready to sell its honey products, he/she has to consider few things such as target market, distribution, marketing channel, packaging and promotion. Initially beekeeper should find among the following options to sell his/her product:

- ❖ friends and family
- ❖ farmers
- ❖ festivals
- ❖ retailers
- ❖ distributors
- ❖ sales outlets
- ❖ local market

Glossary

Abdomen - the third region of the body of a bee enclosing the honey stomach, true stomach, intestine, sting, and reproductive organs.

Absconding swarm - an entire colony of bees that abandons the hive because of disease, wax moth, excessive heat or water, lack of resources, or other reasons.

Acarine disease - The name of the disease caused by the tracheal mite (Acarapis woodi). See Tracheal mite.

Afterswarm - a small swarm which may leave the hive after the first or primary swarm has departed. These afterswarms usually have less bees associated with them than the primary swarm.

American foulbrood - a brood disease of honey bees caused by the spore-forming bacterium, Paenibacillus larvae. The spore stage of the bacterium can remain viable for many years, making is difficult to eliminate the disease.

Apiary - colonies, hives, and other equipment assembled in one location for beekeeping operations; also known as a bee yard.

Apiculture - the science and art of raising honey bees.

Apis mellifera - Genus and species of the Western honey bee originating in Europe and Africa and now located around the world.

Bait hive - A hive or box placed preferably in an elevated location used to attract and hopefully capture swarms.

Bee blower - a motorized blower used as one method to remove bees from honeycombs. Typically frames are not removed from supers prior to using the blower.

Bee bread - a fermented mixture of collected pollen and nectar or honey, deposited in the cells of a comb. Pollen is the primary pollen source for bees and is used especially by the nurse bees to produce royal jelly to feed the young larvae.

Bee brush - a brush or whisk broom used to gently remove bees from combs.

Bee escape - a device used to remove bees from honey supers or buildings by permitting bees to pass one way but preventing their return.

Beehive - a box or receptacle with movable frames, used for housing a colony of bees.

Bee metamorphosis - the three stages through which a bee passes before reaching maturity: egg, larva, and pupa. During the pupal stage, large fat reserves are used to transform both the internal and external anatomy of the bee.

Bee space - 3/8-inch space between combs and hive parts in which bees build no comb or deposit only a small amount of propolis. Bee spaces are used as corridors to move within the hive.

Beeswax - a complex mixture of organic compounds secreted by four pairs of special glands on the worker bee's abdomen and used for building comb. Its melting point is from 143.6 to 147.2 degrees F.

Bee veil - a cloth or form of hat usually made of wire netting to protect the beekeeper's head and neck from stings.

Bee venom - the poison secreted by special glands attached to the stinger of the bee.

Boardman feeder - a device for feeding bees that consists of an inverted jar with an attachment allowing access to the hive entrance.

Bottom board - the floor of a beehive that all the other components build upon.

Brace comb - a small bit of wax built between two combs or frames to fasten them together. Brace comb is also built between a comb and adjacent wood, or between two wooden parts such as top bars.

Braula coeca - the scientific name of a wingless fly commonly known as the bee louse.

Brood - immature bees that not yet emerged from their cells. Brood can be in the form of eggs, larvae, or pupae of different ages.

Brood chamber - the part of the hive in which the brood is reared; may include one or more hive bodies and the combs within.

Burr comb - a bit of wax built upon a comb or upon a wooden part in a hive but not connected to any other part.

Capped brood - pupae whose cells have been sealed with a porous cover by mature bees to isolate them during their nonfeeding pupal period; also called sealed brood.

Cappings - a thin layer of wax used to cover the full cells of honey. This layer of wax is sliced from the surface of a honey-filled comb.

Castes - a term used to describe social insects of the same species and sex that differ in morphology or behavior. In honey bees there are two castes, workers and queens. The drones are a different sex and therefore not included.

Cell - the hexagonal compartment of comb built by honeybees.

Chilled brood - Bee larvae and pupae that have died from exposure to cold. This typically occurs in spring when the colony is expanding rapidly and on cold nights there aren't enough bees to keep the brood warm.

Chunk honey - honey cut from frames and placed in jars along with liquid honey.

Clarifying - removing visible foreign material from honey or wax to increase its purity.

Clarifying Tank - any tank or holding vessel that is use to temporarily store honey while the wax and other material separate from the honey.

Cluster - a large group of bees hanging together, one upon another.

Colony - all the worker bees, drones, queen, and developing brood living together in one hive or other dwelling.

Comb - a mass of six-sided cells made by honey bees in which brood is reared and honey and pollen are stored; composed of two layers united at their bases.

Comb foundation - a commercially made structure consisting of thin sheets of beeswax with the cell bases of worker cells embossed on both sides in the same manner as they are produced naturally by honey bees.

Comb honey - honey produced and sold in the comb. It is produced either by cutting the comb from the frame or when the comb is built in special frames which allow for its easy removal.

Creamed honey - honey which has crystallized under controlled conditions to produce a tiny crystal and a smooth texture. Often a starter or seed is used to help control the crystallization.

Crimp-wired foundation - comb foundation which crimp wire is embedded vertically during the manufacturing of the foundation. The wire increases the strength of the foundation.

Cross-pollination - the transfer of pollen from an anther of one plant to the stigma of a different plant of the same species. Honeybees are excellent pollinators.

Crystallization - the formation of sugar crystals in honey. Syn. Granulation

Cut-comb honey - comb honey cut into various sizes, the edges drained, and the pieces wrapped or packed individually

Dextrose - one of the two principal sugars found in honey; forms crystals during granulation. Also known as glucose.

Dividing - spltting a colony to form two or more colonies.

Division board feeder - a wooden or plastic compartment which is hung in a hive like a frame and contains feed for bees.

Double screen - a wooden frame with two layers of wire screen to separate two colonies within the same hive, one above the other. An entrance is cut on the upper side and placed to the rear of the hive for the upper colony.

Drawn combs - cells which have been built out by honey bees from foundation in a frame.

Drifting of bees - the failure of bees to return to their own hive in an apiary containing many colonies. Young bees tend to drift more than older bees, and bees from small colonies tend to drift into larger colonies.

Drone - the male honey bee

Drone comb - comb measuring about four cells per linear inch that is used for drone rearing and honey storage.

Drone layer - a queen who is incapable of fertilizing eggs. As a result all brood produced is drones.

Dysentery - a condition of adult bees characterized by severe diarrhea and usually caused by starvation, low-quality food,, confinement due to poor weather conditions, or nosema infection.

European foulbrood - an infectious disease which only affects the brood of honey bees and is caused by the bacteria Streptococcus pluton.

Extracted honey - honey removed from the comb.

Extractor - a machine which removes honey from the cells of comb by centrifugal force.

Fermentation - the process of yeast utilizing sugar as a food, and as a byproduct, produce alcohol. Honey typically does not have enough moisture for fermentation to occur.

Fertile queen - a queen, which has been inseminated, naturally or artificially, and can lay fertilized eggs.

Field bees - worker bees generally two to three weeks old that work to collect nectar, pollen, water, and propolis for the colony.

Follower board - a thin board the size of a frame that can be inserting into a hive to reduce the space available to the bees. This is done to help smaller colonies that may have trouble keeping the brood nest warm.

Frame - a piece of equipment made of either wood or plastic designed to hold the honey comb.

Fructose - the predominant simple sugar found in honey.

Fume board - a rectangular cover the size of a super which has an absorbent material on the underside. A chemical is placed on the material to drive the bees out of supers for honey removal.

Fumigilin-B - an antibiotic used in the prevention and suppression of nosema disease.

Grafting - removing a worker larva from its cell and placing it in a queen cup in order to have it reared into a queen.

Grafting tool - a needle or probe designed for transferring larvae from worker cells to a queen cells.

Granulation - the formation of sugar crystals in honey which may cause it to turn solid.

Hive - the structure used by bees for a home.

Hive body - a wooden box which encloses the frames and is usually used as a brood chamber.

Hive stand - a structure that supports the hive.

Hive tool - a metal device used to open hives, pry frames apart, and scrape wax and propolis from the hive parts.

Honey - a sweet viscid material produced by bees from the nectar of flowers, composed largely of a mixture of sugars dissolved in about 17 percent water. It contains small amounts of mineral matter, vitamins, proteins, and enzymes.

Honeydew - a sweet liquid excreted by aphids, leaflhoppers, and some scale insects that is collected by bees, especially in the absence of a good source of nectar.

Honey house - building used for extracting honey and storing equipment.

Honey stomach - a specially designed organ in the abdomen of the honey bee used for carrying nectar, honey, or water.

Increase - to add to the number of colonies, usually by dividing those on hand.

Inner cover - a lightweight cover used under a standard telescoping cover on a beehive.

Instrumental insemination - the introduction of drone spermatozoa into the genital organs of a virgin queen by means of special instruments.

Invertase - an enzyme produced by the honey bee which helps to transform sucrose to dextrose and levulose.

Larva (plural, larvae) - the second stage of bee metamorphosis; a white, legless, grublike insect.

Laying worker - a worker which lays unfertilized eggs, producing only drones, usually in colonies that are hopelessly queenless.

Mating flight - the flight taken by a virgin queen while she mates in the air with several drones.

Mead - honey wine.

Migratory beekeeping - the moving of colonies of bees from one locality to another during a single season to take advantage of two or more honey flows.

Nectar - a sweet and often fragrant liquid secreted by the nectaries of plants for attracting animals. Nectar is the raw product of honey.

Nectar flow - a time when nectar is plentiful and bees produce and store surplus honey.

Nectar guide - color marks on flowers believed to direct insects to nectar sources.

Nectaries - the glands of plants which secrete nectar, located within the flower or on other portions of the plant (extrafloral nectaries).

Nosema - a disease of the adult honey bee caused by the protozoan Nosema apis. The microbe destroys the gut of the bee and severe infections result in malnutrition and dysentery.

Nucleus - a hive of bees which consists of fewer frames than a typical hive and may be smaller in size. A nucleus usually consists of two to five frames of comb and used primarily for starting new colonies or rearing or storing queens; also called and commonly referred to a nuc.

Nurse bees - young bees, three to ten days old, which feed and take care of developing brood.

Observation hive - a hive made largely of glass or clear plastic to allow for the observation of bees at work.

Package bees - a quantity of adult bees (2 to 5 pounds), with or without a queen, contained in a screened shipping cage with a food source.

PDB (Paradichlorobenzene) - crystals used to fumigate stored combs against wax moth.

Pheromones - chemical substances secreted from glands and used as a means of communication. Honey bees secrete many different pheromones.

Play flight - short flight taken in front of or near the hive to acquaint young bees with their immediate surroundings.

Pollen - the male reproductive cell bodies produced by anthers of flowers. It is collected and used by honey bees as their source of protein.

Pollen basket - a flattened depression surrounded by curved hairs, located on the outer surface of a bee's hind legs and adapted for carrying pollen to the hive.

Pollen substitute - any material such as soybean flour, powdered skim milk, brewer's yeast, or a mixture of these used in place of pollen as a source of protein to stimulate brood rearing. Typically feed to a hive in early spring to encourage colony expansion.

Pollen supplement - a mixture of pollen and pollen substitutes used to stimulate brood rearing typically in early spring to encourage colony expansion.

Pollen trap - a device for removing pollen loads from the pollen baskets of incoming bees.

Pollination - the transfer of pollen from the anthers to the stigrna of flowers.

Primary swarm - the first swarm to leave the parent colony, usually with the old queen (see secondary swarm).

Propolis - sap or resinous materials collected from trees or plants by bees and used to strengthen the comb and to seal cracks; also called bee glue.

Pupa - the third stage in the development of the honey bee, during which it changes (pupates) from a larva to an adult bee.

Queen - a female bee with a fully developed reproductive system, and she is larger and longer than a worker bee.

Queen cage - a small cage in which a queen and three to five worker bees are confined for shipping and introduction into a colony.

Queen cell - a special elongated cell in which the queen is reared. It is above an inch or more long and hangs down from the comb in a vertical position.

Queen clipping - removing a portion of one or both front wings of a queen to prevent her from flying.

Queen excluder - metal or plastic device with spaces that permit the passage of workers but restrict the movement of drones and queens to a specific part of the hive.

Robbing - stealing of nectar, or honey, by bees from other colonies which happens more often during a nectar dearth.

Royal jelly - a highly nutritious glandular secretion of young bees, used to feed the queen and young brood.

Sacbrood - a viral disease which affects the larva of honey bees.

Scout bees - worker bees searching for a new source of pollen, nectar, propolis, water, or a new home for a swarm of bees.

Secondary swarm - a smaller swarm which may occur after the primary swarm has occurred.

Skep - a beehive made of twisted straw without movable frames.

Slatted rack - a wooden rack that fits between the bottom board and hive body. Bees make better use of the lower brood chamber with increased brood rearing. Congestion at the front entrance is reduced which can also reduce the swarming tendency.

Slumgum - the refuse from melted comb and cappings after the wax has been rendered or removed.

Smoker - a device in which materials are slowly burned to produce smoke (not flames) which is used to subdue bees. It is important to use a material that produces a cool smoke as not to harm the bees.

Solar wax melter - a glass-covered insulated box used to melt wax from combs and cappings by the heat of the sun.

Spur embedder - a handheld device used for embedding wires into foundation with the purpose of reinforcing the foundation.

Stinger - the modified structure of a worker honey bee used as a weapon of offense. Honey bees have a barbed stinger which stays embedded in the recipient of sting cause the bee to later die.

Streptococcus pluton - bacteria that cause European foulbrood.

Sucrose - principal sugar found in nectar.

Super - any hive body, or smaller box, used for the storage of surplus honey which the beekeeper will harvest. Normally it is placed over or above the brood chamber. Betterbee offers shallow, medium, and deep supers.

Supersedure - the natural replacement of an established queen by a newly reared queen in the same hive.

Surplus honey - honey removed from the hive which exceeds that needed by bees for their own use.

Swarm - a large number of worker bees, drones, and usually the old queen that leaves the parent colony to establish a new colony.

Swarming - the natural process of propagating a colony of honey bees.

Swarm cell - queen cells usually found on the bottom of the combs before swarming.

Terramycin - an antibiotic used to control American and European foulbrood.

Uncapping knife - a knife used to shave or remove the cappings from combs of sealed honey prior to extraction. These can be heated by steam or electricity.

Uniting - combining two or more colonies to form one larger colony.

Virgin queen - a queen which is not mated.

Wax glands - glands that secrete beeswax, which are in pairs on the underside of the last four abdominal segments.

Wax moth - larvae of the moth Golleria mellonclia, which can seriously damage brood and empty combs.

Winter cluster - a ball-like arrangement of adult bees within the hive during winter.

Worker bee - a female bee whose reproductive organs are undeveloped. The majority of the honey bees are worker bees and they do all the work in the colony except for laying fertile eggs.

Worker comb - comb measuring about five cells to the inch, in which workers are reared and honey and pollen are stored.

Bibliography

Accorti, M., Piazza, M.G., Persano Oddo, L. and Sabatini, A.G. (1986) Schede di caratterizzazione delle principali qualita' di miele italiano. Apicoltura, 2: appendi

Accorti, M. 1992. [The influence of the environment on the behaviour and biology of bees in environmental monitoring]. In Proc. of Symp. on "The honeybee as monitoring insect for agricultural pollution1~, 1992, Florcncc 41 53

Adam, Brother 1953. Mead. Bee World, 34(8): 149-156

Adams, F. 1939. The genuine works of Hippocrates. Williams & Wilkins, Baltimore, U.S.A.

Adjare, S.O. 1984. The golden insect, A handbook on beekeeping for beginners. IT Publications, Russell Press Ltd., Nottingham, UK, 104 pp.

Adjare, S.O. 1990. Beekeeping in Africa. FAO Agricultural Series, Bulletin 68/6, Rome, Italy, 130 pp.

Aeppler, C.W. (1922) Tremendous growth force. Gleaning in bee culture, 50: 69-73 Afifi, E.A., Khattab, M.M., El-Berry, A.A., and Abdel-Gawaad, A.A. (1989) Effect of royal jelly on guinea-pig growh. In Proceedings of the Fourth International Conference on

Apiculture in Tropical Climates. Cairo, Egypt, 6-10 November 1988. IBRA, London, U.K.

Ahmed, S.M., Gupta, M.R. and Bhavanagary, M.M. 1976. Stabilization of pyrethrins for prolonged residual toxicity Part II. Development of new formulations. Pyretlirum Post, 13 (4): 119-121

Aleksandrova, L.I. and 8 others 1972. Rosin and beeswax composition (polish). USSR Patent No.329813

Alfa-Laval 1988. Centri-therm: ultra-short-time evaporator for heat sensitive liquids. Lund. Sweden, 8 pp.

Alford, D.V. 1975. Bumble bees. Davis-Poynter, London, UK.

American Bee Journal 1982. Cosmetic natural waxes are not melting away. Amer. Bee I., 122 (12): 822-823

American Bee Journal 1993. U.S. honey, beeswax and pollen prices. Amer. Bee J., 133 (4): 235

Ammon, R. and Zoch, E. (1957) Zur Biochemie des Futtersaftes der Bienenko~mgin. Arzneimitt. Forsch. 7: 699-702

Andrich, G., Fiorentini, R. and Consiglieri, A. 1987. [Characteristics of some samples of propolis from the Ligurian Coast] Caratteristiche di alcuni tipi di propoli della Riviera Ligure. Citta delle Api, (28): 30-31, 34, 35, 37, 38.

Anonimous. 1981. Infant botulism. FDA Drug Bull. 11:11-12

Apimondia 1975a. A valuable product of beekeeping: propolis. Researches and views on its composition, properties and therapeutic value. Apimondia Publishing House, Bucharest, Rumania, 167 pp.

Apimondia 1975b. The hive products: food health and beauty. Proc. of Intern. Symp. on Apitherapy. Apimondia Publishing House, Bucharest, Romania, 154 pp.

Apimondia 1976. Apitherapy today. Apimondia Publishing House, Bucharest, Romania, 105 pp.

Apimondia 1977. New apitherapy research. Proc. of Second Intern. Symp. on Apitherapy. Apimondia Publishing House, Bucharest, Romania, 360 pp.

Apimondia 1978 A remarkable hive product: propolis. Apimondia Publishing House, Bucharest, 250 pp.

Anon 1965. Productions du docteur Anon. French Patent No.1,396,582. In: Chem. Abstr. 63: 6782g (1965).

Armon, P.J. 1980. The use of honey in the treatment of infected wounds. Tropical Doctor, 10: 91.

Arora, D.D. and Kual, K.K. 1973. Feeding practices during the first five years among central Indian communities. Indian Journal of Pediatrics, 40: 203-216.

Asencot, M. and Lensky, Y. (1975) Des larves femelles d'abeille mellif~re, nourries avec de la gelee d' ouvrie'res supplementee, naissent des reines adultes. XXV Congr. Int. Apic. (Grenoble), Apimodia, Bucharest, 271-278.

Asis, M. 1979. El propoleo, un valioso producto apicola. Ministeno de Agricultura, Havana, Cuba, 124 pp.

Asis, M. 1989. [Propolis: the purple gold of honeybees.] Centro de Informaci6n y Documentaci6n Agropecuano, Havana, Cuba, 255 pp.

Ask-Upmark, E. 1967. Prostatitis and its treatment. Acta Med. Scand., 181: 355-357

Aubert, S. and Gonnet, M. 1983. Mesure de Ia couleur des miels. Apidologie, 14 (2): 105-118

Aureli, P., Hatheway, C.L., Fenicia, L. and Ferrini, A.M. 1986. Prime segnalazaioni di casi di botulismo infantile in Italia. Ann. 1st. Super. Sanita', 22: 855-858

Bailey, R.C. 1989. The Efe: archers of the African rain forest. Nat. Geographic, 176: 664-668

Banks, B.E.C. and Shipolini R.A. 1986. Chemistry and pharmacology of honeybee venom. In "Venoms of the Hymenoptera", T. Piek (ed.), Academic Press, London, chpt. 7, 329-416

Bankova, V. and 7 others. 1988. [On the chemical composition of some propolis fractions with antiviral action.] Acta Microbiologica Bulgarica, 23: 52-57

Bankova, V. and 4 others. 1991. A study on the origin of Bulgarian propolis. Apiacta, 26(1): 13-17

Bankova, V., Dyulgerov, A., Popov, S., Evstatieva, L. Kuleva L., Purb 0. and Zamjansan, Z. 1992. Propolis produced in Bulgaria and Mongolia: Phenolic compounds and plant origin. Apidologie, 23 :79-85

Bankova, V. Dyalgerov, A., Popov, S. and Marekov, N.L. 1987. [A GC/MS study of the propolis phenolic constituents.] Z. f. Naturforschung, 42:147-151

Bansal, R.D., Ghosh, B.N., Bhardwaj, U.D. and Joshi, S.C. 1973. Infant feeding and weaning practices at Simla-Hills, Himachal Pradesh. Indian I. Medical Research, 61 (12): 1869-1875

Barral, G. and D. 1992. Propolis ointment with Vaseline. Pers. communication.

Bee Well, 1992 and 1993. The Quarterly Newsletter of the American Apitherapy Society (many case histories and literature reviews) Vol.2 and 3. (see Annex for address)

Bennett, H. 1970. The new cosmetic formulary. Chemical Publ. Comp., New York, 150 pp.

Benkova, M., Boroskova', Z., Dubaj, J. and Sze'chenyi, S. 1989. The immunomodulative effect of propolis preparations on guinea pigs with experimental ascaridosis. Helminthologia 26(2): 163-172

Benson, G.G., Hemingway, S.R. and Leach, F.N. 1978. Coi~p9sition of the wrappings of an ancient Egyptian mummy. J. Pharmacy and Pharmacology, 30 (supplement), 78

Benson, K. 1984. Cleaning and handling pollen. Amer. Bee J. 124: 301-305

Benton, A.W. and Morse, R.A. 1968. Venom toxicity and proteins of the genus Apis. J. Apic. Res., 7 (3): 113-118

Benton, A.W., Morse, R.A. and Stewart, J.D. 1963. Venom collection from honeybees. Science, 142 : 228-230

Bera, L.N., Adam, V.E. and Costea, D. 1971. Solid paste for wear-resistance enhancement of cutting tools. Romania Patent, 51103, 2 pp.

Bergman, A. Yanai, J., Weiss, I., Bell, D. and Menachem, P.D. 1983. Acceleration of wound healing by topical application of honey. An animal model. The American J. of Surgery, 145 : 374-376

Berthold, R. 1993. Beeswax crafting. Wicwas Press, Cheshire, Conn., U.S.A., 128 pp. Berthold. R. 1988a. A new concept in mead making. Amer. Bee J., 128: 820-824 Berthold, R. 1988b. A delicious way to increase sales, honey and fruit, Glean. Bee Cult., 116 (7): 408-410

Bezborodov, V.G. 1968. [Technological properties of slips for the hot casting of glass articles under pressure.] Steklo Keram., 25 (7): 24-27

Bhandari, N.R. and Patel, G.P. 1973. Dietary and feeding habits of infants in various socio-economic groups. Indian Pediatrics, 10 (4): 233-238

Bianchi, E.M. 1990. Control de calidad de la lmiel y de la cera. FAO Agricultural Series, Bulletin No.68/3, Rome, Italy, 69 pp.

Bianchi, E.M. 1991. Preparacidn de Tintura, Extracto Blando, Pomada 0 Unguento, Jabon y otros productos a base de propoleo. CEDIA: 1-30

Birshtein, V.Y., Tul'chinskii, V.M. and Troitskii A.V. 1976. [A study of organic components in ancient Central Asian and Crimean wall paintings.] Vestnik Moskosvkogo Universiteta. 31(3): 33-38

Birshtein, V.Y. and Tul'chinskii, V.M. 1977. [Determination of beeswax and some impurities by infrared spectroscopy.] Khimiya Prirodnykh Sodinenil. (2): 271-275

Blum, M.S., Novak A.F. and Taber III, 5. 1959. 10-Hydroxy-decenoic acid, an antibiotic found in royal jelly. Science, 130 : 452-453

Blum, M.S., Jones, T.H., Rinderer, T.E. and Sylvester, H.A. 1988. Oxygenated compounds in beeswax: Identification and possible significance. Comp. Biochem. and Physiol., 91(3): 581-583

Bogdanov, 5. 1989. Determination of pinocembrin in honey using HPLC. I. Apic. Res., 28 (1): 55-57

Bonomi, A. 1959. [Observations and remarks on the effect of royal jelly on the increase in weight, the haematological picture, the development of male gonads and the demobromatological composition of the flesh of chickens]. 1st. Zootecn. Gen., Univ. Parma, H.H: VI Convengo Salute: 18 pp

Bonomi, A., Marletto F. and Bianchi, M. 1976. [Use of propolis in the food of laying hens.] Revista di Avicultura, 45 (4): 43-55

Bonomi, A. (1983) Acquisizioni in tema di composizione chimica e di attivita' biologica della pappa reale. Apitalia, 10 (15): 7-13.

Borgia, M., Sepe, N., Brancato, V., Simone, P., Costa, G. and Borgia, R. (1984) Efficacia e tollerabilita' di un preparato a base di miele, pappa reale e ginseng in un gruppo di pazienti affette da tubercolosi cronica. Clinica Dietologica, 11 443-447.

Brand, H.M. 1989. Modified beeswax and a process for the modification of beeswax. European Patent Application No. EP 319 062, 21 pp.

Braines, L.N. 1959. Royal jelly I. Inform. Bull. Inst. Pchelovodstva, 31 pp (with various articles)

Braines, L.N. 1960. Royal jelly II. Inform. Bull. Inst. Pchelovodstva, 40 pp.

Braines, L.N. 1962. Royal jelly III. Inform. Bull. Inst. Pchelovodstva, 40

Broadman, J. 1962. Bee venom: the natural curative for arthritis and rheumatism. Putman, New York, 220

Bromenshenk, J.J. and 3 others 1985. Pollution Monitoring of Puget Sound with honeybees. Science, NY, 227 : 632-634

Brown, D. and Kosikowski, F.B. 1990. How to make honey yoghurt. American Dairy Review (4)

Brown, R.H. 1981. Beeswax. Bee Books New & Old, Burrowbridge, UK, 74 pp. (2nd edition in 1989)

Brumfitt, W., Hamilton-Miller, J.M.T. and Franklin, 1.1990. Antibiotic activity of natural products: 1. Propolis. Microbios 62:19-22

Budavari, S. (ed.) 1989. The Merck Index. Merck & Co., Rahway, NJ.

Büll, R. 1959-1977. Vom Wachs: Hoechster Beitra~ge zur Kenntnis der Wachse. (Of waxes:...) Hoechst AG, Frankfurt, Germany, vol.1 in 12 parts, 1097 pp.

Burgett, M. 1990. Bakuti - a Nepalese culinary preparation of giant honeybee brood. The Food Insects Newsletter, 3 (3): 1-2

Burlando, F. 1978. [About the therapeutic action of honey on burn wounds.] Minerva dermatolog. 113: 699-706

Buscigho, J.A. 1988. ssAnti-inflammatgory topical compositions containing lidocaine and diphenhydramine [and propolis]. USA Patent No.4 748 002, 5 pp.

Canadian Honeybee Research Association (CHRA), 1988. The propolis trap. CHRA, B.C. Newsletter, June, 2-3

Carli, H.O. De., Cornejo, L.G. and Maljar, L. (1975) Aterosclerosis expermental. Ministeno de Asuntos Agrarios, Buenos Aires, Argentina, (1975), 63 pp.

Cartland, B. 1970. The magic of honey. Corgi Books, London, UK, 160 pp.

Cavanagh, D., Beazley, 1. and Ostapowicz, F. 1970. Radical operation for carcinoma of the vulva. A new approach to wound healing. The 1. of Obstet. & Gynaec. BC, 77 (11)1037-1040

Chauvin, R., Deftomont, C., Louveaux, 1. and Verge', 1.1952. Sur une substance presente dans le pollen, qui s'oppose au developpement de certaines bacteries. Compt. Rend. Soc. Biol. 146: 645

Chauvin, R. and Louveaux, 1. (1956) Etdue macroscopique et microscopique de lagelee royale. L'apiculteur.

Chauvin, R. Action physiologique et therapeutique des produits de la ruche. In Traite' de biologie de l'abeille. Paris, France, Masson et Cie, (1968) Tomme III, 116-1154.

Chernyak, N.F. 1973. On synergistic effect of propolis and some anti-bacterial drugs. Antibiotiki, 18 : 259-261

Chirife, 1., Scarmato, G. and Herszage, L. 1982. Scientific basis for use of granulated sugar in treatment of infected wounds. The Lancet, : 560-561

Cho, Y.T. 1977. Studies on royal jelly and abnormal cholesterol and triglycerides. Amer. Bee 1., 117 : 36-38

Cho, H., Toni, M. and Kanamori, T. 1988. [Deodorants controlling mouth odour.] Japanese Patent No. JP 63 264 516 [88 264 516], 6 pp.

CHRA. 1988. The propolis trap. Newsletter, Canadian Honeybee Research Assocition of British Columbia, June, 2-3

Clauss, B. 1982. Beekeeping handbook. Ministry of Agriculture, Gaborone, Botswana.

Clauss, B. and R. 1991. Zambian Beekeeping Handbook. Mission Press, Ndola, Zambia, 108 pp.

Codex Alimentarius Commission 1989. Codex standards for sugards (honey). Supplement 2 to Codex Alimentarius Volume III. Food and Agriculture Organization of the United Nations and World Health Organization, Rome.

Codex Alimentarius 1994. Honey. 2nd Edition, FAO/WHO, Vol.11: 21-24

Codex Alimentarius 1995. Standards for honey. 2nd Education, FAO/WHO Vol.13 (in preparation).

Coggshall, W. and Morse, R.A. 1984. Beeswax: production, harvesting, processing and products. Wicwas Press, Ithaca, NY, USA, 192 pp.

Colangelo, M. 1980. Combo-packed yoghurt and granola gives convenience a healthy image. Dairy Field 163 (10): 95-96, 98, 100.

Cohn, M.E. and 3 others 1986. La qualite' des miels du commerce. Cah. Nut. Diet., 21 (3): 219-222

Contari, G. 1987. [Process for the propolis extract preparation]. Apicolt. Mod., 78 :147-150

Contessi, A. (1990) Le api, biologia, allevamento, prodotti. Edagricole, Bologna

Coprean, D. and 4 others. 1986. [Effect of standardized propolis extract in the experimentally intoxicated liver of rats.] Clujul Medical, 59(4): 333-337

Cortopassi-Laurino, M. and Gelli, D.S. 1991. Analyse pollinique, propriete's physicochimiques et action antibacterienne des miels d'abeilles africanisees Apis mellifera et de Meliponine's du Bresil. Apidologie 22: 61-73

Cosmetics and Toiletries. 1992. Various formulae. Cosmetics and Toiletries 107(4): 96, 99

Costantini, F. and Ricciardelli d'Albore, G. 1971. Pollen as an additive to the chicken diet. Proc. 23rd International Apicultural Congress, Apimondia, p.539-542.

Couture, H. and Guzzi, D. 1989. Candle making using beeswax. Leaflet from Trop. Beekeeping Newsletter, IBRA, Cardiff, UK, 3 pp.

Crane, E. 1975. Honey: A comprehensive survey. Heinemann (in coop. with IBRA), London, U.K. 608 pp.

Crane, B. 1979. Honey in relation to infant botulism. Bee World, 60(4): 152-154

Crane, B. 1980. A book of honey. Oxford University Press, Oxford, U.K., 198 pp.

Crane, E. 1984. Bees, honey and pollen as indicators of metals in the environment. Bee World, 65 (1): 47-49

Crane, E. 1990. Bees and beekeeping: Science, Practice and World Resources. Cornstock Publ., Ithaca, NY., USA. 593 pp

Cuellar Cuellar, A. and Rojas Hernandez, N.M. 1987. [Chemical components of Cuban propolis, 1.] Revista Cubana de Farmacologia, 21(3): 365-372

Cue llar Cue llar, A., Rojas Hernandez, N.M. and Martinez Perez, J. 1990. p[New antimicrobial structure from propolis collected in Cuba.] Revista Cubana de Farmacia, 24 (1): 51-58

Csuka, J., Baumgartner, J. and Dubay, J. 1978. [The effect of royal jelly on some reproductive characters of Japanese quail.] Zivocisna Vyroba, 23 (5): 395-400

Dadant & Sons 1975. The hive and the honeybee. Dadant & Sons, Hamilton, Illinois, USA, 740 pp. (see Graham, 1992, for newer edition)

Dadant, 1992. The hive and the honeybee. Dadant & Sons, Hamilton, Ill., USA, 1324 pp. (see also Graham, 1992)

Daharu, P.A. and Sporns, P. 1984. Phenol residue levels in honey. J. Apic. Res., 23 (2): 110-113

Dany, B. 1988. Selbstgemachtes aus Bienenprodukten. Ehrenwirth Verlag, Mu~nchen, 174 pp.

De Belfever, B. (1958) La gelee royale des abeilles. Maloine, Paris.

Debuyser, E. 1984. La propolis. These pour dipl6me de docteur en Pharmacie. Fac. Pharmacie, Univ. Nantes, France, 34 pp.

DeNavarre, M.G. 1962. The chemistry and manufacture of cosmetics. Vol.11 - Cosmetic Materials. D. Van Nostrand Comp., Princeton, NJ., USA.

Denis, L.J. 1966. Chronic prostatitis. Acta Urol. Belg., 34: 49-55

Derevici, A. and Petrescu, A. (1965) Experimental studies in vitro and in vivo on the virulicidal action of royal jelly. Lucr. Stunt. Stat. cent. Sen. Apic. (1965), 5:135-143.

Destrem, H. (1956) Experimentation de la gelee royale d'abeille en pratique geriatrique (134 cas). Rev. Franc. Geront, 3.

Destrem, H. (1981) La gelee royale chez les personnes agees. Revue Fran~aise d'apiculture, numero speciale (Apitherapie 1981)16-18.

Devakumar, C., Baskaran, S. and Mukerjee, 5K. 1986. Isolation of N=triacontanol from Indian beeswax and its effect on dry matter of rice. Indian Journal of Agricultural Sciences, 56(10): 744-747

Diaz Gonzalez, J.A. and Iglesias Perez, H. 1977. Characteristics of Cuban beeswax. Revista Cubana de Farmacia, 11(1): 75-82

Dietz, W., Martin, P.J. and Harcus. J.F. 1976. Bees as a protein supplement for pigs. Can. J. Anim. Sci., 56 (4): 841

Dietz, A. 1975. Nutrition of the adult honeybee. In: The hive and the honeybee. Dadant & Sons, Hamilton, Ill., p.125-156

Dietz, A. and Stevenson, H.R. 1975. The effect of long-term storage on the nutritive value of pollen for brood rearing of honeybees. Amer. Bee J. 115: 476-477, 482

Dietz, A. and Stevenson, H.R. 1980. Influence of long-term storage on the nutritional value of frozen pollen for brood rearing of honeybees. Apidologie, 11:143-151

Dillon, J.C. and Louveaux, J. Pollen et gelee royale. Cah. Nutr. Diet. (1987), 22 (6): 456-465.

Dimov, V., Ivanovska, N., Bankova, V. and Popov, 5. 1992. Immunomodulatory action of propolis: IV. Prophylactic activity against Gram-negative infections and adjuvant effect of the water-soluble derivative. Vaccine, 12 (V): 1-7

Dimov, V., Ivanovska, N., Manolova, N., Bankova, V., Nikolov N. and Popov, 5. 1991. Immunomodulatory action of propolis: Influence on anti-infectious protection and macrophage function. Apidologie, 22:155-162

Donadieu, Y. 1975. Le miel. Maloine Ed., Paris, 36 pp.

Donadieu, Y. 1979. La propolis. Editions Maloine, Paris.

Donadieu, Y. 1980. La gelee royale. Maloine, Paris

Donadieu, Y. 1983. Honey in natural therapeutics. Paris: Maloine Editeur, S.A., 28 (2nd.)

Donadieu, Y. and Marchiset, C. 1984. La cire (wax).. Editions Maloine, Paris, 131 pp.

Doner, L.W., Chia, D. and White, J.W. 1979. Mass spectrometric 13C/12C determinations to distinguish honey and C3 plant syrups from C4 plant syrups (sugar cane and corn) in candied pineapple and papaya. J. Ass. Off. Analyt. Chem., 62 (4): 928-930

Dorato, S. 1987. Water solub le jplant extracts in cosmetics. Cosmetics and Toiletries 102: 70-73

Dotimas, E.M. and Hider, R.C. 1987. Honeybee venom. Bee World, 68 (2): 51-70

Driesche, D. Van 1983. Hand-dipped beeswax candles. Amer. Bee J., 123 (3): 173-176

Dubaj, J. and 7 others. 1988. [Agent for the regeneration of damaged tissue containing pantothenic acid, zinc, and extract of propolis.] Czech Patent No. CS 253 424, 13 pp.

Dubovsky, I. and 6 others. 1988. [Propolis-stabilized vitamin C.] Hungarian Patent No. HU 46 849, 8 pp.

Dumronglert, E. 1983. A follow-up study of chronic wound healing dressing with pure natural honey. J. Nat. Res. Council, Thailand, 15(2): 39-66

Dunn, J.D. 1984. The effect of bee venom on plasma corticosterone levels. Neuroendocrinology Letters, 6 (5): 273-277

Dustmann, J.H. 1979. Zur antibakteriellen Wirkung des Honigs. Apiacta, 14: 7-11

Dustmann, J.H. and Gunst, E. 1982. Inhibins and bacteriostatic action of beebread. Apiacta, 17: 51-54

Dustmann, J.H. and von der Ohe, W. 1988. Radioaktivita~tsmessungen in Honig aus m.edersachsischen Trachtgebieten. Nordwestdeutsche Imkerzeitung, 40 (5): 129-131

Dutta, R.K. 1959. Indian Bee Journal 21:110 as cited in Sharma and Singh, 1983

Dyce, E.J. 1975. Producing finely granulated or creamed honey. In: 1'Honey, a comprehensive survey". by Crane, (ed.), Heinemann, London, U.K. 293-306

Eason, T. 1991. Hand dipping beeswax candles. Amer. Bee J., 131: 617-619

Ebisu, T., Maeda, N., Matsubara, H. and Kato, K. 1988. [Honey powder compositions as additives for foods and pharmaceuticals.] Japanese Patent No.63 157 943 [88 157 943], 3 pp.

Efem, S.E.E. 1988. Clinical observations on the wound healing properties of honey. British J. of Surgery, 75: 679-681

Eischen, F.A. and Dietz, A. 1987. Growth and survival of Galleria mellonella (Lepidoptera: Pyralidae) larvae fed diets containing honeybee-collected plant resins. Ann. Entomol. Soc. Am., 80 (1): 74-77

Eischen, F.A. and Dietz, A. 1990. Improved culture techniques for mass rearing Galleria mellonella (Lepidoptera: Pyralidae). Entomological News, 101(2): 123-128

EI-Banby, M.A. 1987. Honeybees in the Koran and in medicine. Al-Ahram Centre for Translation and Publication, Cairo, Egypt, 205 pp.

El-Banby, M.A. and 4 others. 1989. Healing effect of floral honey and honey from sugar-fed bees on surgical wounds (animal model). In: Proc. 4th Intern. Conf. Apic. Trop. Climates, Cairo, Egypt. 6-10 November 1988. IBR, London, U.K. p.46-49.

El-Sabbagh, H.M., Abdel-Gawad, H.A. and El-Said, Y. 1988. Development and characterization of an oleogineous suppository base. Alexandria Journal of Pharmaceutical Science, 2(1): 80-83

EI-Shahaly, A.A., Mohamed, M.S., El-Zalaki, E.M. and Mohasseb, Z.S. 1978. Formulation, storage possibilities, and chemical composition of ready-to-eat honeytahena paste. Libyan J. Agriculture, 7: 65-72

Enger, C.C. 1976. Impermeable silicone composition. US patent No.3963677, 6 pp. In Chem. Abstr. 85:79451

Fahmy, F.G. and Omar, M.O.M. 1989. Effect of propolis extracts on certain potato viruses. Proc. 4th Intern. Conf. Apic. Trop. Climates, Cairo, Egypt, November 1988: 56-60

Faruga, A., Puchajda, H. and Bobrzecki, J. 1975. [Utilization of bee hive waste in poultry feed]. Zeszyty Naukowe Akad. Rolniczo Techn., W Olsztynie, 142:151-158

Feinberg, S.M. and Gothers. 1940. Oral pollen therapy in ragweed pollinosis. J. Amer. Med. Assoc. 115: 23-29

Feinberg, W. 1983. Lost-wax casting. Intermed. Technol. Publ. London, 74 pp

Fenfenati, L., Sabatini, A.G., Nanetti, A. (1986) Composizione in sali minerali della gelatina reale. Apicoltura, 2:129-143.

Ferber, C.E.M. and Nursten, H.E. 1977. The aroma of beeswax. J. Sci. Fd. Agric., 28: 511-518

Filho, O.M. and Carvalho, A.C.P. de. 1990. Application of propolis to dental sockets and skin wounds. Journal of Nihon University School of Dentistry, 32(1): 4-13

Fotin, A.V. and Gel'medova, N.N. 1981. (Treatment of allergic rhinosinusitis in children using honeybee venom). Vestnik Otorinolaringologii, (4): 42-44

Fox, C. 1992. Cosmetics & Toiletries, 107: 70-88

Franco, T.T. and Kurebayashi, A.K. 1986. [Isolation of the main active constituents of propolis by two-dimensional paper chromatography and spectrophotometry]. Revista do Inst. Adolfo Lutz, 46 (1/2): 81-86

Free, J.B. and S others 1983. Using foraging honeybees to sample an area for trace metals. Environ. Entomol., 9: 9-12

Friedmann, H., Kern, J. and Rust, J.H. 1957. The domestic chick: a substitute for the honey guide as a symbient with cerolytic. Amer.Nat., 91(860): 321-325

Frilli, F., Barbattini, R. and Milani, N. 1989. L'ape, atlante. (The bee atlas.) University of Udine, Dept. of Plant Defenses, Italy and ERSA, 40 pp.

Fuchs, L. 1960. The problem of royal jelly therapy. Medsche Welt, Stuttg., 40: 2119-2121

Fuchs, F.J. 1970. High pressure continuous wire extrusion. Western Electric. Princeton, N.J., USA, 18 pp.

Fuj ii, A., Kobayashi, S., Ishihama, S. Yamamoto, H. and Tamura, T. 1990. Augmentation of wound healing by royal jelly. Japan. J. Pharmacol., 53 (3): 331-337

Furness, C. 1974. Beeswax Candles. British Bee Publications, Gedd., 54 pp

Furness, C. 1977. How to make beeswax candles. British Bee Publications, Geddingtron, U.K., 14 pp.

Furness, C. 1986. Beeswax Candles. British Bee Publications, Geddington, U.K., 20 pp.

Gabrys, J. and 4 others 1986. Free amino acids in bee hive products (propolis) as identified and quantified by gas-liquid chromatography. Pharmac. Research Communications 18 (6): 513-518

Gafar, J., Sacalus, A., David, E. and David, N. 1986. [Treatment of simple pulp gangrene with the apitherapy product "Propolis".] Stomatologie, 33(2): 115-117

Gaider, F.A. 1950. Pchelovodostvo, 37: 55, as cited in Sharma and Singh, 1980.

Gala Books 1971. 2000 down home skills & secret formulas for practically everything. Gala Books, Laguna Beach, CA.

Galuszka, H. 1972. The research on a most effective method of the collection of bee venom by means of electric current. Zoologica Pol., 22 (12): 53-69

Gary, N.E., Ficken, R.W. and Stein, R.C. 1961. Honeybee larvae (Apis mellifera L.) for bird food. Avicult. Mag., 67: 27-32

Gasparri, F. 1983. [Beekeeping and cosmetics. Elements for an evaluation of the use of honey and propolis in cosmetics, criteria for the choice of their application and qualitative analysis methods for the original material and the final product.] Italy, Proc. Conf. Erboristeria e Cosmetico, Bologna. Erboristeria Domani - Quaderni, 30-35

Gayre, G.R. 1948. Wassaill in mazers of mead. Phillmore &.Co., London, 176 pp.

Gayre, G.R. and Papazian, C. 1986. Brewing mead - Wassaill in mazers of mead. Brewers Pubi, Boulder, Co. USA, 200 pp.

Geiskopf, 5. 1984. Putting it up with honey: a natural food canning and processing cookbook. Quicksilver Productions, Ashland, OR, USA, 219 pp.

Gentry, C. and 3 others. 1985. A manual for trainers of small-scale beekeeping development workers. Peace Corps, Washington, D.C., USA, 2nd edition, 407 pp.

Gentry, C. 1988. Small scale beekeeping. Peace Corps, Wash. D.C., USA, 213 pp.

Ghisalberti, E.L. 1979. Propolis: a review. Bee World, 60 (2): 59-84

Gimbel, N.S., Threlkeld, R., Farris, W. (1962) Epithelization in experimental burn blisters. Research in Burns. Am. Inst. Biolog. Sci. (9).

Ginsberg, N.J., Dauer, M. and Slotta, K.H. 1968. Melittim used as a protective agent against X-irradiation. Nature, 220 :1334

Giordani, G. 1961. [Effect of royal jelly on chickens.] Avicoltura 30 (6): 114-120

Giurgea, R. and 4 others. 1987. Biochemical effects of standardized propolis extract (SPE) and of silymarin on the liver of ethyl alcohol intoxicated rats. Agressologie, 28(8): 831-832.

Giugea, R., Rusu, M.A., Popescu, H. and Polinicencu, C. 1989. [The liver-protecting action of standardized propolis extract in ethanol poi;soning of rats.] Clujul Medical, 62(1): 56-59

Gondal, S.M.A. and Hashmi, A.A. 1976. Effect on egg production in Apis cerana indica queens on feeding them with royal jelly or crushed larvae. J. Apic. Res. 15(3/4): 130-132

Gonnet, M. 1973. [Controlled granulation of honey and production of creamed honey.] Revue fran~aise d'apiculture, 309: 209-212

Gonnet, M. 1977. Liquefazione, pastorizzazione e cristallizzazione controllata del miele. In: "Miele: aspetti tecnologici" by Apimondia, Bucharest : 61-66

Gonnet, M. 1985. Miel creme. Review Fran~aise d'Apiculture, 12: 591-593.

Gonnet, M. et Vache, G. 1985. Le gout du miel. U.N.A.F., Paris. 146 pp.

Gonnet, M. 1986. Une bonne pate pour un bon tube. Revue Fran~aise d'Apiculture, (1): 29-30

Gonnet, M. 1986. L'analyse des miels. Description de quelques methodes de contr6le de la qualite'. Bull. Tech. Apicole, 13(1): 17-36.

Gonnet, M., Jeanne, F. and Faucon, A. 1988. L'hydromel. Bull. Tech. Apicole. 15(3): 139-146.

Gorbatenko, A.G. 1971. [Treatment of ulcer patients with a 30% alcohol solution of propolis.] Vrach. Delo, 3: 22-24

Graham, J.M. Ed. 1992. The hive and the honeybee. Dadant & Sons, Hamilton, Illinois, U.S.A. 1324 pp.

Grange, J.M. and Davey, R.W. 1990. Antibacterial lproperties of propolis (bee glue). J. Roy. Soc. Medicine, London, U.K. 83(3): 159-160

Grange, J.M. 1990. Honey and propolis as possible promotors of the healing of ulcers in leprosy (reply letter, comment). Lepr. Rev. 61(2): 195.

Green, A.E. 1988. Wound healing properties of honey. British J. of Surgery, 75 (12): 1278

Greenaway, W., Scaysbrook, T. and Whatley, F.R. 1987. The analysis of bud exudate of Populus X euramericana and of propolis by gas chromatography-mass spectrometry. Proc. Royal Soc. London B, 232: 249-272

Greenaway, W., Scaysbrook, T. and Whatley, F.R. 1988. ~omposition of propolis of Oxfordshire, lU.K. and its relation to poplar bud exudate. Zeitschrift fur Naturforschung, C 43: 301-304

Greenaway, W., May, J., Scaysbrook, T. and Whatley, F.R. 1990a. Identification by gas chromatography - mass spectrometry of 150 compounds in propolis. Zeitschrift fu~r Naturforschung, C, 46:111-121

Greenaway, W., Scaysbrook, T. and Whatley, F.R. 1990b. The composition and plant origins of propolis: a report of work at Oxford. Bee World, 71:107-118

Grohmann, F. and Spitznagel, H.U. 1988. Processing your propolis the right way. Canadian Honeybee Research Association of British Columbia Newsletter, August : 5-6

Grunberger, D. and 7 others. 1988. Preferential cytotoxicity on tumor cells by caffeic acid phenethyl ester isolated from propolis. Experientia, 44: 230-232

Gubicza, A. and Molnar, P. 1987. [Propolis in the rearing of calves.] Magyar Mezo~gazdasa'g, 42(17): 14

Gueorguieva, E. and Vassilev, V. 1990. Traitement de l'ulce're par la propolis. Revue fran~aise d'apiculture, 499: 394-397

Gunnison, A.F. 1966. An improved method for collecting the liquid fraction of bee venom. J.Apic. Res., 5 (1): 33-36

Guralnick, M.W., Mulfinger, L.M. and Benton, A.W. 1986. Collection and standardization of Hymenoptera venoms. Folia Allerg. et Immunol. Clinica, 33 (1): 9-18

Guss, S.B. 1967. Bee larvae as food for caged birds. Amer. Bee J., 107: 62

Haffejee, I.E. and Moosa, A. 1985. Honey in the treatment of infantile gastroenteritis. Brit. Med. J. 290:1866-1867

Halberstadt, K. (1978) Etude comparative de la nourriture larvaire et des glandes nourricieres chez l'abeille domestique sur le plan de l'enzymologie et de la chimie des proteins. Apidologie, 9:143-146.

Hasegawa, M., Saeki, Y. and Sato, Y. 1983. [Artificial rearing of some beneficial insects on drone powder and the possibility of their application.] Honeybee Science, 4:153-156

Hashimoto T., Takeuchi, K., Hara, M. and Akatsuka, K. 1977. Pharmacological study on royal jelly (RJ). 1. Acute and subacute toxicity tests on RJ in mice and rats. Bulletin of the Meiji College of Pharmacy No.7 1-13.

Hashimoto, T., Tori, M., Asakawa, Y. and Wollenweber, E. 1988. Synthesis of two allergenic constituents of propolis and poplar bud excretion. Zeitschrift flir Naturforschung, C 43: 470-472

Hausen, B.M., Wollenweber, E., Senff, H. and Post, B. 1987. Propolis allergy. (I). Origin, properties, usage and literature review. Contact Dermatitis, 17:163-170

Hausen, B.M. and Wollenweber, E. 1988. Propolis allergy: (III) Sensitization studies with minor constituents. Contact Dermatitis, 19: 296-303

Haydak, M.H. 1967. Bee nutrition and pollen substitutes. Apiacta, 1: 3-8

Heggers and 6 others. 1987. Control of burn wound sepsis: a comparison of in vitro topical antimicrobial assays. The J. of Trauma. 28(2): 176-179

Henschler, D. (1954) Hoher Acetylocholingehalt von Bienenfuttersa~ften. Naturwissenschaften, 41:142.

Hepburn, H.R. 1986. Honeybees and Wax: An Experimental Natural History. GFR: Springer-Verlag, Berlin, Gewrmany, 205 pp.

Herbert, E.W. Jr. 1992. Honeybee nutrition. In: The hive and the honeybee, Graham, Ed., :197-233

Hernuss, P. and 6 others 1975. Pollendia~t als Adjuvans der Strahlentherapie gynaekologischer Karzinoma (Pollen diat as adjuvant to radiation therapy of gynaecological carcinoma). Strahlentherapie, 150: 500-506

Hetland, A. 1986. Clostridium botulinum sporer i norskprodusert honning? Norsk. Vet. Artidsskr., 98 (10): 725-727

Higginbottom, J.A. 1974. A guide to wax and dyecraft materials: a comprehensive catalogue of materials for candlemaking, tie-&-die and batik. Candle Makers Supplies, London, U.K., 47 pp.

Hill, R. 1977. Propolis, the natural antibiotic. Thorsons, Wellingborough, U.K.

Hill, K. Hawkes, K., Hurtado, M. and Kaplan H. 1984. Seasonal variance in the diet of Ache hunter-gatherers in eastern Paraguay. Human ecology, 12: 101-135

Hofmann, U., Holst, H. and Schlo~sser, E. 1989. [Studies of the effect of plant protection agents on the susceptibility of grapevines to Plasmonara viticola. 2. Results of an infection trial.] Wein-Wissenschaft 44: 61-65

Holderna, E. and Kedzia, F. 1987. Investigations upon the combined action of propolis and antimycotic drugs on Candida albicans. Herba Polonica, 33(2): 145-151

Hollands, I., Miyares, C., Sigarroa, A. and Perez, A. 1984. [Efficacy of propolis against infection by intestinal Eimeria in rabbits.] Revista Cubana de Ciencias Veterinarias, 15(2): 157-163

Hollands, I., Miyares, C. and Sigarroa, A. 1988. [Comparative analysis of the action of propolis, sulphaquinoxaline and sulphamethazine in rabbits with coccidiosis.] Rivista Cubana de Ciencias Veterinarias, 19(2): 99-104

Hollands, I., Miyares, C. and Pimienta, R. 1988. [Quality ;control of Propolisina (alcoholic extract of propolis) used as a coccidiostat, by means of a biological method.] Revista Cubana de Ciencias Veterinarias, 19(4): 319-326

Ho Shin 1980. (Api-acupuncture). Ho Shin, No.1, 120 pp.

Hocking, B. and Matsumura, F. 1960. Bee brood as food. Bee World, 41(5): 113-120

Howe, S.R., Dimick, P.S., Benton, A.W. (1985) Composition of freshly harvested and commercial royal jelly. J. Apic. Res., 24 (1): 52-61.

Hsiang, H.K. and Elliott, W.B. 1975. Differences in honeybee (Apis mellifera) venom obtained by venom sac extraction and electrical milking. Toxicon 13: 145-148

Hughes, N.B. 1960. Casting composition. US Patent 2,942,995 (1960).

Huhtanen, C.N., Knox, D. and Shimanuki, H. 1981. Incidence of Clostridium botulinuim spores in honey. J. of Food Protection, 44(11): 812-814

Hunt, K.J. and 5 others 1978. A controlled trial of immunotherapy in insect hypersensitivity. New Eng. I. Med., 299:157-161

Hutton, D.J. 1966. Treatment of pressure sores. Nursing Times, 18:1533-1534 limori, Y. 1975. A rust inhibitor (incorporating beeswax). Japanese Patent, 75 15 484, 3 pp. Ikeno, K., Ikeno, T. and Miyazawa, C. 1991. Effects of propolis on dental caries in rats. Caries Research, 25: 347-351

Ikuta, H. 1931. The investigation of Japanese beeswax. Analyst, 56: 430-436

Inoue, T. and Inoue, A. 1964. The world royal jelly industry: present status and future prospects. Bee World, 45 (2): 59-69.

Inoue, T. 1986. The use and utilization of royal jelly and the evaluation of the medical efficacy of royal jelly in Japan. Proceeding sof the XXXth International Congress of Apiculture, Nagoya, 1985, Apimondia, 444-447

Inoue, H. 1988. [Propolis, its chemical constituents and biological activity.] Honeybee Science, 9(3): 115-126

ITDG 1978. Simple methods of candle manufacture. Intermed{ate Technology Development Group, London, U.K., 19 pp

ITC, UNCTAD/GATT 1977. Major markets for honey: Openings for quality supplies from developing countries. ITC Publications, Geneva, Switzerland, 120 pp.

ITC, UNCTAD/GATT 1978. The world market for beeswax: a high value product requiring little investment. ITC Publications, Geneva, 105 pp.

ITC, UNCTAD/GATT 1986. Honey, a study of major markets. ITC Publications, Geneva, 167 pp.

ITC, UNCTAD/GATT. 1986. Report on the markets for selected hive products in the United Kingdom, France and Italy. Prospects and developments (ins French). ITC Publications, Geneva, Switzerland, 35 pp.

Iwamoto, N., Nakano, A. and Imanishi, Y. 1965. Method for filling hard capsules with granular drugs. D.B.R. Patent No.1207047 (1965).

Iwasaki, M. 1990. [Propolis-containing antibiotic ointments for atopic dermatitis treatment.] Japanese Patent No. JP 02 142 734 [90 142 734], 2 pp.

Jacoli, G. (1956) Ricerche sperimentali su alcune proprieta' biologiche della gelatina reale. Apicoltore d'Italia, 23 (9-10): 211-214.

Jean, E. 1956. A process of royal jelly absorption for its incorporation into assimilable substances. Fr. Pat., 1,118,123

Jeanne, F. 1985. La refonte du miel. Bull. Tech. Apicole. 12(1): 33-40

Jeddar, A. and 5 others 1985. The antibacterial action of honey. S. Afric. Med. J., 67: 257-258

Jiang, J. 1986. [Rape oil steroid isolation from beeswax for use as a plant growth stimulator.] Chinese Patent No. CN 85 102 899, 5 pp.

Jindra, M. and Sehnal, F. 1989. Larval growth, food consumption, and utilization of dietary protein and energy in Galleria mellonella. J. Insect Physiol., 35 (9): 729-724

Johansen, C. 1955. Bee-collected pollen for artificial pollination of apples. Amer. Bee J., 95: 352-353

Johansson, T.S.K., Johansson, M.P. (1958) Royal Jelly II. Bee World, 39: 254-264, 277-286.

Jolly, V.G. 1977. Propolis violin varnish. Strad, 88: 713-719

Jolly, V.G. 1978. Propolis varnish for violins. Bee World, 59 (4): 158-161

Jones, CL. 1977. The balance of beeswax retained in synthetics. Manuf. Chemist & Aerosol News, 48 (3): 46-50

Kaal, J. 1991 Natural medicine from honeybees. 93 pp. (available from Bees and Development)

Kaatz, H. 1986. Heisser Honig? Nach Tschernobyl: Auch der Honig ist radioaktiv belastet. Allg. dtsch. Imkerzeitung, 20 (7): 222-224

Kandil, A., El-Banby, M., Abdel-Wahed, K. Abou Sehly, G. and Ezzat, N. 1987a. Healing effect of true floral and false nonfloral honey on medical wounds. J. of Drug Research (Egypt), 17 (1-2): 71-75

Kandil, A., El-Bandy, M., Abdel-Wahed, K., Abdel-Gawaad, M. and Fayez, M. 1987b. Curative properties of true floral and false non-floral honeys on induced gastric ulcers. J. Drug Research Egypt, 17(1-2): 103-106

Kandil, A., El-Banby, M.A., Abdel-Wahed, K., Abdel-Gawaad, M. and Fayez, M. 1989. Curative properties of floral honey and honey from sugar-fed bees on induced gastric ulcers. Proc. 4th Intern. Conf. Apic. Trop. Clim., 68-69

Karaali, A., Meydanoglu, F. and Eke, D. (1988) Studies on cp--mpositon, freeze drying and storage of Turkish royal jelly. J. Apic. Res., 27 (3): 182-185.

Karimova, Z.H. and Rodionova E.I. 1975. Propolis in the treatment of lung tuberculosis. In: A valuable product of beekeeping: propolis. Apimondia 1975. Edition Apimondia, Bucharest, Romania.

Katsilambros, N.L., Philippides, P., Touliatou, A., Georgakopoulos, K., Kofotzouli, L. Frangaki, D., Siskoudis, P., Marangos, M. and Sfikakis, P. 1988. Metabolic effects of honey (alone or combined with other foods) in type II diabetics. Acta Diabetologica Latina, 25 (3): 197-203

Kaul, 5. 1967. Erfahrungen mit Bienenhom.glu~sungen in der Allgemeinpraxis. Physik. Med. u. Rehab., 8: 265-266

Kedzia, A. 1986. [Effect of ethanol extract of propolis (EEP) on anaerobic bacteria.] Herba Polonica, 32(1): 53-58

Kedzia, B, and Holderna, E. 1986. [Investigations on the combined action of antibiotics and propolis on Staphylococcus aureus.] Herba Polonica, 32(3/4): 187-195

Kedzia, B., Iwaszkiewicz, J. and Geppert, B. 1988. [Pharmacological investigations on ethanolic extract of propolis.] Herba Polonica, 34(4): 243-253

Kel'man, I.M. 1960. Application of bee venom in sanatorium conditions. Pchelovodstvo, 37 (3): 52-54

Kerkvliet, J.D. 1981. Analysis of a toxic rhododendron honey. J. Apic. Res. 20(4): 249-253

Khattab, M.M., Radwan, A.A. and Afifi, E.A. (1989) Physiological effect of royal jelly on female reproductive capacity in rabbits. In Proceedings of the Fourth International Conference on Apiculture in Tropical Climates. Cairo, Egypt, 6-10 November 1988. IBRA, London, U.K.

Killion, E.E. 1992. The production of comb and bulk comb honey. In: The hive and the honeybee by Graham, Dadant & Sons, Ill., U.S.A. 705-722

Kim, C.M. 1989. Bee venom therapy for arthritis. Rheumatology, 41(3): 67-72

Kime, R.W., McLellan, M.R. and Lee, C.Y. 1991. An improved method of mead production. Am. Bee J. 131(6): 394-395

Klein, K. 1991. Kukui and Macadamia nut oils. Cosmetic applications. Cosmetics & Toiletries, 106: 87-90

Knopf, E. and Ogait, A. 1961. [Why is German propolis no longer suitable for varnishing stringed instruments?] Instrumentenbau Z., 5: 152-160

Kondo, T. 1979. Microencapsulation. Techno Inc., Tokyo, 119 pp

König, B. and Dustmann, J.H. 1985. Fortschritte der Celler Untersuchungen zur antivirotischen Aktivita~t von Propolis. Apidologie, 16 (3): 228-230

Köm.g, B. 1988. (News about propolis: Anti-tumor effect and allergies from caffeic acid derivates). Heilkunst, 101(10): 453-456

König, B. and Dustmann, J.H. 1989. Tree resins, bees, and antiviral chemotherapy. Animal Research and Development, 29: 21-42

Kramer, K.J., Childs, C.N., Spiers, R.D. and Jacobs, R.M. (1982) Purification of insulin-like peptides from insects haemolynph and royal jelly. Insect Biochemistry 12 (1):91-98.

Kramer, K.J. Tager, H.S., Childs, C.N. and Spiers, R.D. (1977) Insulin-like hypoglycemic and immunological activities in honeybee royal jelly. Journal of Insect Physiology (1977), 23 (2): 293-295.

Krause, M.V. and Mahan, L.K. 1979. Food, nutrition and diet therapy. W.B. Saunders Co., Philadelphia, USA, 963 pp.

Krell, R. 1990. A simple modification for the centrifugal extraction of combs harvested from frameless hives. (similar to Krell, 1991). Appr. Technology, 17 (3)

Krell, R. 1991. Centrifugal honey extraction in frameless-hive beekeeping. Beekeeping & Development, 19: 6-7

Krell, R. 1992. A simple method for reducing the moisture content of tropical honeys. Proc. 5th Intern. Conf. Apic. in Tropical Climates, Trinidad and Tobago, 38-43.

Krell, R., Persano Oddo, L. and Ricciardelli D'Albore, G. 1988. The influence of harvesting and processing methods on honey quality in Zambia and Malawi. Proc. of the 4th Internat. Conf. on Apiculture in Tropical Climates, Cairo: 268-273

Krochmal, C. (Undated) Natural cosmetics, from beehive to herb garden. Selfpublished, Asheville, USA, 35 pp.

Krochmal, C. 1973. A guide to natural cosmetics. Times Book Comp., Quadrangle, New York, 227 pp.

Krochmal, C. 1985. Hive cosmetics. Glean. Bee Cult., 113 (10): 527-529

Krochmal, C. 1991. Hive cosmetics. Amer. Bee J. , 131: 573-576

Krzyzynski, K. 1988. [Varnish from dammar resin and/or wax [beeswax] for paintings and art restoration.] French Patent No. FR 2 607 513, 5 pp.

Kühn, H. 1960. Detection and identification of waxes, including Punic wax, by infra-red spectrography. Stud. Conserv., 5 (2): 71-81

Kurstjens, S.P., McClain, E. and Hepburn, H.R. 1990. The proteins of beeswax. Naturwissenschaften, 77 (1): 34-35

Labochev, S.V., Marenkov, G.M. and Salnikov, V.1. 1958. Pchjelovodstvo 35: 50, as cited in Sharma and Singh (1980).

Lagrange, V. 1991. Ultrafiltration of Honey. Am. Bee J. 131(6): 453-455, 458

Laidlaw, H.H., Jr. 1979. Contemporary Queen Rearing. Dadant and Sons, Inc., Hamilton, Illinois, USA.

Laidlaw, H.H., Jr. 1992. Production of queens and package bees. In: The hive and the honeybee. ed. J.M. Graham. Dadant and Sons, Hamilton, Illinois, USA, 989-1042.

Lanyon, W.E. and Laynon, V.H. 1969. A technique for rearing passerine birds from the egg. Living bird, 8: 81-93

La Torre, A., Guccione, M. and Imbroglini G. 1990. (Preliminary observations on the action of propolis based preparations against Botrvtis cinerea Pers. on Strawberries. Apicoltura, 6:169-177

Lawrence, W.V. 1986. Infant botulism and its relationship to honey: a review. Am. Bee J., 126: 484-486

Lavie, P. 1968. Proprie te's antibacteriennes et action physiologique des produits de Ia ruche et des abeilles. In vol.3 of: Chauvin, 1968, p.1-115.

Lavie, P. Les substances antibiotiques dans la colonie d'abeille. In Traite' de biologie de l'abeille, Paris, France, Masson et Cie, (1968) Tome III, 1115.

Lee, C.Y. and Kime, R.W. 1984. The use of honey for clari;fying apple juice. J. Apic. Res. 23(1): 45-49

Lee, M.H. and Lee, Y.H. 1987. [Preparation and evaluation of yellow beeswax matrixes and nalidixic acid.] Seoul University Journal of Pharmaceutical Sciences, 12: 33-43

Lejeune, B. and 5 others 1984. [Propolis: extracts and uses in shampoos and lotions.] Parfums, Cosmetiques, Aromes, (56): 65-68

Lejeune, B., Pourrat, A. and Dehmouche, H. 1988. Propolis utilisation en dermocosmetologie. Parfums, Cosmetiques, Aromes, 8~2: 73-77

Lercker, G., Capella, P., Conte, L.S., Ruinji, F. and Giordani, G. (1981) Components of royal jelly. I. Identification of organic acids. Lipids 16: 912-919

Lercker, G., Capella, P., Conte, L.S., Ruini, F. and Giordani, G. (1982) Components of royal jelly: II. The lipid fraction, hydrocarbons and sterolds. J. Apic. Res. 21(3):178-184.

Lercker, G., Vecchi, M.A., Sabatini, A.G. and Nanetti, A. 1984. Controllo chimicoanalitico della gelatina reale. Riv. Merceol. 23 (1): 83-94.

Lercker, G., Savioli, S., Vecchi, M.A., Sabatini, A.G., Nanetti, A. and Piana, L. (1986) Carbohydrate Determination of Royal Jelly by High Resolution Gas Chromatography (HRGC). Food Chemistry, 19: 255-264.

Lercker, G., Caboni, M.F., Vecchi, M.A., Sabatini, A.G. and Nanetti, A. (1992) Caratterizzazione dei principali costituenti della gelatina reale. Apicoltura 8:11-21.

Lilley, W. 1983. Bee miners join British Columbia gold hunt. Amer. Bee J., 123 (9): 635-637

Lloyd, G. 1957. Polish and Shine: Recipes of Women's Institute members and their ancestors. WI Books Ltd., London, UK, 40 pp.

Louveaux, J., Maurizio, A., and Vorwohl, G. 1978. Methods of melissopalynology. Bee World. 59(4): 139-157

Lücke, H. 1935. Wundbehandlung mit Honig und Lebertran. (Wound treatment with honey and codliver oil.) Dtsch. Mediz. Wochenschrift, 41:14-17

Lüpke, J. 1980. Bienenhonig in Trockenform. Kakao u. Zucker., 32 (2): 43

Ltihrs, B. 1935. Honig in a~usserlicher Anwendung. Ztschr. f. Veterina~rkunde, 47: 57-58

Lukoschus, F.S. and Keularts, J.L.W. 1968. [A further function of the mandibular gland of worker honeybees (Apis mellifera L.): Production of a substance inhibiting pollen germination]. Ztschr. f. Bienenforsch., 9 (8): 333-343

Lupachev, V.F. 1963. (Pharmacology of Apilac). Farmak. Toks., 26 (3): 333-338

Majno, G. 1975. The healing hand: man and wound in the ancient world. Harvard Univ. Press, Cambridge, MA., USA, 571 pp.

Makarov, F.D. 1972. (Propolis treatment of ulcer disease and pyloroduodenitis). Vrach. Delo, 4: 93-96

Maksimova-Todorova, V. and 7 others. 1985. [Antiviral effects of some fractions isolated from propolis.] Acta Microbiologica Bulgarica 17: 79-85

Malak, B. 1964. Method of preparing a varnish for furniture. Polish Pat. No.: 48,688. In: Chem. Abstr. 63: 8612b (1965)

Malossi, C., Grandi, F. (1956) Osservazioni sulla gelatina reale nell'alimentazione degli immaturi. Atti del 10 convegno nazionale per lo studio dell'applicazione dei prodotti delle api nel campo medico-biologico, Bologna, Italia, 130-133.

Manolova, N. and 5 others. 1987. [Immunobiological effect of propolis. I. Effect on cellular immunity.] Acta Microbiologica B ulgarica 21: 76-81 (Abstract)

Marchenay, P. 1977. La propolis. Marchenay Philippe, Paris.

Marcos, A. 1991. [Extraction and treatment of pollen]. Abeille de France et l'Apiculteur, 756: 27-32

Marko, P., Pechan, J. and Vittek, J. Some phosphorous compounds in royal jelly. Nature (1964) 202:188-189.

Marletto, F. 1983. Caratteristiche della propoli in funzione dell'origine florale e dell'utilizzazione da parte delle api. Apicolt. Mod., 74 (5)187-191

Marston, N. and Campbell, B. 1973. Comparison of nine diets for rearing Galleria mellonella. Ann. Entomol. Soc. Am., 66:132-136

Marston, N. and Brown, B. 1974. Constituents in diets for Galleria melonella. J. Econ. Entomol. 67(4): 497-500

Martinetti, R. and Caracristi, C. (1956) Azione eccitometabolica della gelatina reale nell'uomo. Atti del l0 convegno nazionale per lo studio dell'applicazione dei prodotti delle api nel campo-medico-biologico, Bologna, Italia, 139-144.

Matsuka, M., Watanabe, M. and Nujima, K. 1982. Longevity and oviposition of vedalia beetles on artificial diets. Environ. Entomol., 11: 816-819

Maxwell, H. 1987. A small-scale honey drying system. Amer. Bee J., 127 (4): 284-286 Melampy, R.M. and Jones, D.B. (1939) Chemical composition and vitamin of royal jelly. Proc. Soc. Expt. Biol. Med. 41: 382-388.

Melampy, R.M. and Stanley, A.J. 1940. Alleged gonadotropic effect of royal jelly. Sience 91: 457-458.

Meresta, L. and Meresta, T. 1985/1986. Antibacterial activity of flavonoid compounds of propolis, occurring in flora in Poland. Bulletin of the Veterinary Institute in Pulawy, 28-29(1-4): 61-63

Meresta, T. and Meresta, L. 1988. [Sensitivity of Bacillas larvae to an extract of propolis in vitro.] Medycyna Weterynayjna, 44(3): 169-170

Meresta, L., Meresta, T., Burdzinski, J. and Chmurzunski, P.1989. [Treatment of mastitis in cows using an extract of propolis.] Medycyna Weterynaryjna, 45(7): 392-3905

Meyer, P. 1977. Nuclear energy and the environment. Kerntechnik, 19 (1): 9-13

Mikhailov, A. C. 1950. The application of medicated honey to eye diseases. Pchelovodstvo, 2:117-118

Milena, L., Leifertova, I. and Baloun, I. 1989. [Fungistatic effect of propolis.] Folia Pharm. Univ. Carol, 13: 29-44

Miller, R.M. 1974. Figure sculpture in wax and plaster. David & Charles, 175 pp.

Millet-Clerc, J., Simeray, J., Michel, D. and Chaumont, J.P. 1986. (Antifungal properties of propolis against fungi causing mycoses). Bulletin de Ia Socie'te' Fran~aise de Mycologie Me'dicale, 15(2): 517-521.

Millet-Clerc, J., Michel, D., Simeray, J. and Chaumont, J.P. 1987. [Preliminary study of the antifungal properties of propolis compared with some commercial products.] Plantes Me'dicinales et Phytothe'rapie, 21(1): 3-7

Minrath, W.R. 1957. Van Nostrand's practical formulary. D. Van Nostrand Company, Inc., Princeton, NJ.

Mitev, B. 1971. (Collection of bee venom using a weak electric current - its effect on the condition and the performance of the colony). Zhivot. Nauki, 8 (1): 103-108

Mizuno, M., linuma, M. and Kato, H. 1987. [Useful ingredients and biological activity of propolis.] Fragrance Journal, 15(2): 20-28

Mizuno, M. 1989a. [Propolis- or its extract-containing resin compositions.] Japanese Patent No. JP Ol 245 058 [89 245 058], 5 pp.

Mizuno, M. 1989b. [Food packaging materials containing propolis[as a preservative].] Japanese Patent No. JP Ol 243 974 [89 243 974], 5 pp.

Mladenov, 5. 1972. Mierea Si therapia cu miere. Editura Ceres, Bucharest, 264

Mlagan, V. and Sulimanovic, D. 1982. Action of propolis solutions on Bacillus larvae. Apiacta, 17:16-20

Molan, P.C., Smith, I.M. and Reid, G.M. 1988. A comparison of the antibacterial activity of some New Zealand honeys. J. Apic. Res., 27 (4): 252-256

Molan, P.C., Allen, K.L., Tan, S.T. and Wilkins, A.L. 1989. Identification of components responsible for the antibacterial activity of Manuka and Viper's Bugloss honeys. Ann. Conf. New Zealand Inst. fo Chem., Paper No.1

Monteverdi, T. and Reitano, S. 1972. [Eutrophic effect of a "natural food" (queen honeybee larvae) in a group of psychiatric patients.] Minerva Dietologica 12(4): 133-144

Morgan, J.F., Tolnai, S. and Townsend, G.F. (1960) Studies on the in vitro antitumor activity of fatty acides: ii. saturated dicarboxylic acids. Can. J. Biochem. Physiol. 38: 597-603.

Morishita, H., Nonogaki, S., Suzuki, A. and Morioka, M. 1978. Ink compositions. Japanese Kokai, 53-9607, 2 pp. In: Chem. Abstr. 89: 26242t (1978)

Morse, R.A. 1964. Honey wine and how to make it. Glean. Bee Cult., 92 (6):338-343 Morse, R.A. and Benton, A.W. 1964a. Notes on venom collection from honeybees. Bee World. 45 (4): 141-143

Morse, R.A. and Benton, A.W. 1964b. Mass collection of bee venom. Glean. Bee Cult., 92 (1): 42-45,54

Morse, R.A. 1965. The effect of light on comb construction by honeybees. J. Apic. Res. 4(1): 23-29

Morse, R.A. and Steinkraus, K.H. 1975. Wines from the fermentation of honey. In:

Honey: a comprehensive survey, by Crame, ed.: 392-407

Morse, R.A. 1978. Comb honey production. Wicwas Press, Ithaca, NY, USA, 128 pp.

Morse, R.A. 1980. Making mead (Honey wine). Wicwas Press, Cheshire, Conn., USA

Mraz, C. 1982. Bee venom for arthritis - an update. Amer. Bee J., 122:121-123

Mraz, C. 1983. Methods of collecting bee venom and its utilization. Apiacta, 18: 33-34, 54

Mucsi, I. and 4 others. 1989. [Drug for treating muscle hypoplasia in piglets, based on propolis extract and vitamins.] Hungarian Patent No. HU 49 809, 13 pp.

Müller, E. 1938. [The venom production of honeybees] Verhandlungen VII. Kongress Entomol., p.1857-1864

Muniategui, S., Simal, J., H;uidobro, J.F. and Garcia, M.C. 1989. [Study of the fatty acids in bee-collected pollen.] Grasas y Aceites, 40(2): 81-86

Munoz, L.G. 1989. Prevention of legs' affections in ovines using propolis. Apiacta 24(3: 80-81

Munro, J.A. 1943. The viscosity and thixotrophy of honey. J. Econ. Ent. 35(5): 769-777 Nagy, E., Pa'pay, V., Litkei, G. and Dinya, Z. 1985. Investigation of the chemical constituents, particularly the flavonoid components, of propolis and populi gemma by the GC/MS method. Studies in Organic Chemistry - Flavonoids and Bioflavonoids. 23: 223-232

Nagy, M. and S others 1989. Constituents of propolis of Czechoslovak origin. V. Chemical Papers, 42 (5): 691-696

Nahmias, F. 1981. [Cure yourself with honey.] Curatevi con il miele. De Vecchi Editore, Milano, 125 pp.

Nahmias, F. 1983. La miel cura y sana. Translation from Italian of Nahmias, 1981, Editorial De Vecchi, Barcelona, 96 pp.

Nakajin, S., Okiyama, K., Yamashita, S., Akiyama, Y. and Shinoda, M. (1982) Effect of royal jelly on experimental hypercholesterolemia in rabbits. Shoyakugalku Zasshi (1982) 36 (1): 65-69.

Nakamura, T. (1986) Quality standards of royal jelly for medical use. proceedings of the XXXth International Congress of Apiculture, Nagoya, 1985 Apimondia (1986) 462-464.

Nakanishi, K., Oltz, E.M. and Grunberger, D. 1989. Preparation of caffeic acid esters as anti-inflammatory agents and cell growth inhibitors. International Patent No. WO 89 000 851, 32 pp.

Nakrashevich, V.F., Stoikov, S.A. and Bronnikov, V.1. 1988. Harvesting pollen from combs. Pchelovodstvo, 10: 29-30

Narayana, N. 1970. Studies in Indian honeys and bees waxes. Centr. Res. Inst., Poona, India, 13 pp.

Nardi, M. (1986) Legislazione ed indicazioni di mercato della pappa reale. Atti del Convegno "Apinfiera" Faenza (Ra) 13-17.

Naum lyorish 1974. Bees and people. Mir. Publisher, Moscow, 212 pp.

Negri, G. 1979. Nuovo erbano figurato. Hoepli, Milano, Italy, 459 pp.

Neumann, D., Go~tze, G. and Binus, W. 1986. [Clinical study of the testing of the inhibition of plaque and gingivitis by propolis.] Stomatologie der DDR, 3612): 677-681

Neychev, H. and 7 others 1988. Immunomodulatory action of propolis. II. Effect of water-soluble fraction on influenza infection in mice. Acta Microbiologica Bulgarica, 23:58-62

Nikolov, N., Marekov, N., Bankova, V., Popov, S., Ignatova, R. and Vladimirova, I. 1987. Method for the preparation of a water-soluble derivative of propolis. Bulg. Patent Appl. 79903/28,05

Nilsson, S., Praglowski, J. and Nilsson, L. 1977. Atlas of airborne pollen grains and spores in Northern Europe. Ljungfoeretagen, Oerebro, Sweden, 159 pp.

NozNick, P.P. and Li, Kwoh H. 1967. Cigarette filters. US Patent #m#!#m#)% (1967). Nye, M.J., Shuel, R.W., Dixon, S.E. Gluconic acid in the food of larval honeybees. J. Apic. Res. (1973)12 (1): 9-15.

Obreg6n Fuentes, A.M. and Rojas Herna'ndez, N. 1990. [Antimicrobial action of alcoholic extracts of propolis.] Revists Cubana de Farmacia, 24(1): 34-44 (Es)

Ochi, T. 1981. A new method to collect propolis. Honeybee Science, 2 (1): 16

Ohta, N. 1983. [Experiences with api-acupuncture]. Honeybee Science, 4(1): 21-24

Okonenko, L.B. and 6 others. 1988. [Vitamin E. and propolis as antioxidants in the presence of excessive polyunsaturated fatty acids.] Voprosy Pitaniya No.4, 68-70

Okonenko, L.B. 1988, [Salmonella infections and propolis.] Zdravookhr. kaz. 1: 55-57

Okada, 1.1971. [An artificial rearing of a coccinellid beetle Harmonia axyridis Pallas, using a diet of larvae and pupae of the worker honeybee.] Collecting Breed., Tokyo, 33(10): 229-236

Okada, I. and Matsuka, M. 1973. Artificial rearing of Harmonia axyridis on pulverized drone honeybee brood. Environ. Entomol., 2: 301-302

Okonenko, L.B. 1986. [Propolis as an inhibitor of lipid free-radical oxidation in salmonellosis.], Voprosy Meditsinskoi Khimii, 32(3): 45-48

Olarin, T., Palos, E. and Olarin, A.I. 1989. Treatment of giardiosis [Giardia infection] with propolis tincture. Proc. 31st Intern. Congr. Apic. Warsaw, Poland, August 1987: 470-473

Olstrom, J.M. 1983. Dried honey. Am. Bee J. 123: 656-659

Omar, M.O.M. 1989. Some characteristics of propolis from Upper Egypt. Proceedings of the Fourth International Conference on Apiculture in Tropical Climates, Cairo, Egypt, 6-10 November 1988, 88-92

Olschki, L. S. (ed.). 1977. Ceroplasty in science and the arts Proceedings of the First International Congress, Florence 3-7 June 1975. Florence, Italy, 728 pp.

Otani, H., Oyama, J. and Tokita, F. 1985. Polyacrylamide gel electrophoretic and immunochemical properties of proteins in royal jelly. Jap. J. Diary Food Sci., 34:21-25.

Paillon, F. 1960. La fabrication des produits alimentaires au miel. Girardot et Cie., Paris, 238 pp.

Palmer, D.J. 1961. Extraction of bee venom for research. Bee World 42(9): 225-226

Pan, A.T.F. and Matsumoto, K. 1975. [An air-permeable waterproofing composition containing paraffin, fatty acids and beeswax.] Patent Applic. No.2256365, 11 pp. French. In: Chem. Abstr. 83: 195180c.

Pa'pay, V. and 4 others. 1985. Isolated compounds from Hungarian propolis and populi gemma. Studies in Organic Chemistry - Flavonoids and Bioflavonoids, 25: 233-240

Papay, V. and 3 others 1987. Chemical and pharmacological study of propolis from various locations. Acta Pharmac. Hung., 57:143-151

Paterson, P.D. 1975. The traditional making of honey bee throughout tropical Africa. In:

Honey: a comprehensive survey; by Crane (ed.): 405-407

Paysen, J. 1987. Wet honey, a method for drying honey on a commercial scale. Amer. Bee J., 127 (4): 273-282

Pectihacker, H. and Huettinger, E. 1986. [Harvesting propolis with high pressure air.] Bienenvater, 107 (5): 160-161

Pence, R.J. 1981. Methods for producing and bio-assaying intact honeybee venom for medical use. Amer. Bee J., 121(10): 726-731

Pepeljnjak, S., Jalsenjak, I. and Maysinger, D. 1981. Influence of microencapsulated propolis extract on Bacillus subtilis strain IP-5832. Acta Pharmaceutica Jugoslavica, 31(1): 27-32

Persano, A.L. 1987. Hidromieles: Historia, recetas y methodos para su elaboraci6n. Editorial Hemisfeno Sur, S.A., Buenos Aires, Argentina, 152 pp.

Persano Oddo, L., Krell, R. and Ricciardelli D'Albore, G. 1988. Contribution to the identification of the geographical and botanical origin of honeys produced in Zambia and Malawi. Apicoltura, 4:113-138

Petri, G., Lemberkovics, E. and Foldvari, M. 1988. Examination of differences between propolis (bee glue) produced from different floral environments. In Flavours and Fragrances: a world perspective (eds. Lawrence, B.M., Mookherjee, B.D., Willis, B.J.). Elsevier Sci. Publ., Amsterdam, 439-446

Phadke, R.P., Nair, KS. and Nandekar, KU. 1969. Indian beeswaxes. 1. Their physicochemical constants. Indian Bee J., 31(2): 52-55

Phadke, R.P.and Nair, K.S.. 1970. Standards for Indian honeys and beewaxes. Indian Bee J. 32(3/4): 68-74

Phadke, R.P., Nair, K.S. abd Babdedjarm K.U. 1971. Indian bees-waxes. 2. The nature of their chemical constituents. Indian Bee J. 33 (1/2): 3-5

Phadke, R.P. and Nair, K.S. 1973. Studies on Indian honeys. 5. Distinguishing characteristics of the apiary honey from the wild variety. Indian Bee J. 35 (1-4): 36-39

Piek, T. ed. 1986. Venoms of the Hymenoptera. Academic Press, London, U.K.

Platt, J.L. and Ellis, J.R.B. 1985. Removing water from hon~y at ambient pressure. US Patent, 4,536,973: 6 pp.

Polishchuk, L.A. and Denisova, L.M. 1970. Carbon paper. USSR Pat. No.275732

Popescu, H., Polinigencu, C., Atansiu, P. and Predescu, E. 1985. Antiherpes ointment. Patent application, Rom. RO 86,003 (Cl. A61 K9/06) 30 Jan.1985, Appl. 108,265, 24 Jul.1982, 2p. (in Chem. Abstr. 1985, 103, 26, # 220838q)

Popravko, S.A. 1977. [Chemical composition of propolis and its standardization.] Pchelovodstvo, 97 (8): 21-23

Poryardin, V.T. 1960. Pchelovodostva, 37: 55, (as cited in Sharma and Singh, 1980).

Posey, D.A. 1978. Ethnoentomological survey of Amerind groups in lowland Latin America. Florida Entomologist, 61(4): 225-229

Pourtallier, J., Taliercio, Y. and Mussot, J.M. (1970) Controle de Ia qualite' de Ia gele'e royale. Revue bibliographique des travaux sur la composition, les proprie'te's et les utilisations de la gele'e royale. Bulletin Apicole No. 2:145-160.

Pourtallier, J., Davico, R. and Rognone, M.C. (1990) Les analyses dans le controle de Ia purete' de Ia gele'e royale. Abeille de France et l'Apiculteur 753: 405-407.

PRC. 1990. Japanese honey products, by Pacific Research Consulting Inc., (PRC), New Food Products in Japan, 15(1), 15(2) abd 15(3)m as cuted ub Katrabgem 1991.

Princeton, N.J. 1970. High pressure continuous wire extrusion. Western Electric, 18

Proserpio, G. and Martelli, A. 1982a. Propolis: Its use in cosmetics. Part I. Il Prodotto Chimico, 9:15-21

Proserpio, G. and Martelli, A. 1982b. Propolis: Its use in cosmetics. Part II. II Prodotto Chimico, 10: 25-30

Proserpio, G. 1981 L'ape cosmetica. Erboristeria Domani Libri, Milano, Italy. 140 pp. Proserpio, G. 1988. Impieghi cosmetici dei prodotti dell'alveare. Erboristeria Domani, 36-41

Prosperi, P., Ragazzini, F. and Francalancia, L. (1956) Sull'impiego terapeutico della pappa reale delle api negli stati di denutrizione della prima infanzia. Atti del l0 convegno nazion~e per lo studio deU' ap~icazione dei prodotti d~e api nel campo medicobiologico, Bologna, Italia 134-136.

Prosperi, P. and Ragazzini, F. (1956) Applicazioni cliniche della gelée royale in campo pediatrico. Riv. Clin. Pediat. 58 (3): 319-332.

Przybylaski, J. and Scheller, 5. 1985. [Early results in treatment of Legg-Calve-Perthes illness with intra-articular injections of aqueous solutions of propolis.] Zeitschrift fur Orethopa~due, 123(2): 163-167

Pyke, E.J. 1973. A biographical dictionary of wax modellers. Oxford University Press, Oxford, U.K. 216 pp.

Quadri, 5. (1956) Applicazione della gelatina reale nelle distrofie della prima infanzia. Clin. Pediat., Modena 38 (9): 686-690.

Quagho, P., Messi, P. and Fabio, A. 1988. [An investigation about the presence of bacteria of the genus Clostridium in honey samples.] L'Igiene Moderna, 15 (3): 486-496

Qureshi, N. and Tamhane, D V.1985. Production of mead by immobilized whole cells of ~haromces cerevisiae. Appl. Microbiol. & Biotechn. 21: 280-281

Radu-Tudorache, G., Oita, N., Luca, A. and Hritcu, V. 1978. [Observations concerning the biostimulant effect of royal jelly on young calves.] Cercetari Agronomice in Moldova, 2:131-133

Rai, B.K., Allicock, P. and Delph G. 1977. Honeybee and injection of monocrotophos into coconut palm. Indian Bee Journal, 39:13-14

Rekka, E. and Kourounakis, P.1990. Antioxidant activity of and interlenkin production affected by honeybee venom. Arzneimittelforschung 40(8): 912-913

Rembold, H. (1965) Biological active sustance in royal jelly. Vitamins and hormones 23:359-382.

Rembold, H., Lackner, B. and Geistbeck, I. (1974) The chemical basis of honeybee, Apis mellifera, caste formation. Partial purification of queen bee determinator from royal jelly. J. Insect Physiol. 20: 307-314.

Revathy, V. and Banerji, S.A. 1980. A preliminary study of antibacterial properties of Indian honey. Ind. J. Biochem. & Biophys. 17(Suppl. No.242): 62

Ribot, E. 1960. (Beeswax and sperm oil as separation agents substitute for liquid paraffin). Zuckeru. S~sswarenw., 13: 190, 196

Ricciardelli D'Albore, G. and Battaglini Bernardini, M. 1978. Origine geographique de la gelée royale. Apidologie, 9 (1): 1-17

Riccardelli D'Albore, G. and Persano Oddo, L. 1978. Flora apistica Italiana. Istituto per Ia Zoologia Agraria, Firenze, Italy, 286 pp.

Ridi, M.S. el, Mofty;, A. el, Khalifa, K. and Solimen, L. 1960. Gonado tropic hormones in pollen grains of the date palm. Z. Naturf. 156(1): 45-49

Rigby, J. and Hepburn, R. 1981. Beeswax candles from a silicone rubber mould. South African Bee Journal, 53 (4): 11-13

Rodriguez L6pez, C. 1985. Determinaci6n espectro-fotome'trica del color de las mieles. Vida apic., 16: 24-29

Rojas Herna'ndez, N.M. and Cue'tara Bernal, K. de la. 1990. [Antibiotic effect of propolis against strains of Staphylococcus aureus of human clinical origin.] Revista Cubana de Farmacia, 24(1): 45-50

Rombauer, I.S. and Rombauer Becker, M. 1975. Joy of cooking. Bobbs-Merrill Company, New York, 915 pp.

Ross, P.B. 1990. The effects of propolis fractions on cells in tissue culture. M.Phil. Thesis, University of Wales College of Cardiff, U.K. xii + 193 pp.

Roubik, D. (ed.) 1995. Pollination in Tropical Agriculture. FAO Bulletin 68/... (in preparation).

RSFSR 1977. (Standard for propolis). RST RSFSR :317-77

Ruttner, F. ed. 1983 Queen Rearing. Apimondia Publishing House, Bucharest, Romania.

Ryan, J.K., Jelen, P. and Sauer, W.C. 1983. Alkaline extraction of protein from spent honeybees. J. Food Science, 48: 886-888, 896.

Sagawa, M. 1983 Success and failure in api-acupuncture. Honeybee Science, 4 (1): 27-28
Sakai, T., Sasaki, M. and Suzuki, T. 1978. [Stability of pulverized drone honeybee brood as a diet for mass-rearing of predacious insects: effects of storing conditions and gamma-irradiation on the development of a lady beetle Harmonia ~~~y~ffidis.] Bull. Faculty Agriuclture, Tamagawa University 18:69-76

Salajan, G. 1970. Inst. Agron. "Dr. Petru Groza" Luc. Stut. Ser. Zootech. 26:165 as cited in Stanley and Linskens, 1974, p.113 and Schmidt and Buchmann, 1992.

Salama, A., Mogawer, H.H. and El-Tohamy, M. 1977 Royal jelly a revelation or a fable. Egyptian Journal of Veterinary Science 14 (2): 95-102.

Salem, S.N. 1981. Honey regimen in gastrointestinal disorders. Bull. Islamic Med. 1: 358-362

Salmon, R.E. and Szabo, T.I. 1981. Dried bee meal as a feedstuff for growing turkeys. Can. J. Anim. Sci., 61: 965-968

Sanyal, S.K. and Roy, S.K. 1967. Low temperature behaviour of temporary corrosion protectives. Labdev J. Sci. Technol., 5(3): 216-220

Sangalli, A. 1990. La propoli. L'Ape Nostra Amica, 12 (4): 16-25

Sargant, J. 1971. Two hundred years of wax modelling: a history of Madame Tussaud's. Central Assoc. of Bee-Keepers, Ilford, U.K. 10 pp.

Sasaki, M., Tsuruta, T. and Asada, 5. (1987) Role of physical property of royal jelly in queen differentiation of honeybee. In Chemistry and biology of social insects (edited by Eder, J., Rembold, H.). Munich, German Federal Republic, Verlag J. Papemy 306-307.

Sato, T. 1977. (Pollens for skin cosmetics.) Japanese Kokai, Patent application No. 90634/1977, 3 pp.

Savina, K.A. and Romanov, F.T. 1956. Propolis as a medicinal remedy. Pchelovodstvo 33(8): 59-60

Serkedjieva, J. 1992. Anti-influenza virus effect of some propolis constituents and their analogues (esters of substituted cinnamic esters). J. Natural Products, 55(3): 294-302

Scales, J.T. and Winter, G.D. 1961. The adhesion of wound dressings. An experimental study. In: Wound healing, D. Slome, (ed.)

Scheller, S. and 6 others. 1989a. The ability of ethanolic extract of propolis EEP to protect mice against gamma irradiation. Zeitschrift fur Naturforschung, C, 44:1049-1052

Scheller, S. and 7 others. 1989b. [Trials of immunoregulation in patients with chronic bronchitis.] Immunologia Polska 14(3/4): 204-305

Scheller, S. and 5 others. 1989c. [Immunization trials in two cases of alveolitis fibroticans with decreasing conductivity of the immune system: effect of ethanol extract of propolis (EEP), Esberitox N. and a calcium-magnesium preparation.] Heilkunst, 102(6): 249-255

Scheller, S. and 7 others 1989d. Trace elements in propolis and in its ethanolic extract (EEP) as determined by neutron activation analysis. Z. f. Naturf., 44:170-172

Scheller, S. and 5 others. 1989e. Antitumoral property of ethanolic extract of propolis in mice-bearing Ehrlich carcinoma, as compared to bleomycin. Zeitschrift fur Naturforschung. C. 44:1063-1065

Scheller, S. and 4 others. 1990. Free radical scavenging by ethanol extract of propolis. Int. J. Radiation Biol. 57(3): 461-465

Schmidt, J.O. 1992. Allergy to venomous insects. In: The hive and the honeybee. J.M. Graham, ed. Dadant & Sons, Hamilton, Illinois, 1209-1269.

Schmidt, J.O. and Buchmann, S.L. 1992. Other products of the hive. In: The hive and the honeybee J.M. Graham, ed. Dadant & Sons, Hamilton, Illinois, USA. 927-988

Schumacher, M.J., Schmidt, J.O. and Egen, W.B. 1989. Lethality of "killer" bee stings. Nature, 337: 413

Shambaugh, P., Worthington, V. and Herbert, J.H. 1990. Differential effects of honey, sucrose, and fructose on blood sugar levels. J. Manipul. Physiol. Therapeutics, 13 (6): 322-325

Sharma, H.C. and Singh, O.P. 1983. Medicinal properties of some lesser known but important bee products. Proc. 2nd Int. Conf. Apiculture in Trop. Climates, IBRA, New Delhi, March 1980. 694-702

Shawer, M.B., Ah, S.M., Abdellatif, M.A. and El-Refai, A.A. 1987. Biochemical studies of bee-collected pollen in Egypt. 2. Fatty acides and non-saponifiables. J. Apic. Res., 26(2): 133-136

Shen, Z. and Zhao, Y. 1986. [Plant growth substances in beeswax. 1. Gibberellin-like substances.] Zhiwu Shenglixue Tongxun No.2, 31-32

Shinoda, M. and 5 others 1978. [Biochemical studies on vasodilative factor in royal jelly.] Yakugaku Zassii, 98:139-145

Shipman, W.H. and Cole, L.J. 1967. Increased resistance of mice to X-irradiation after injection of bee venom. Nature, 215: 311-312

Shipolini, R.A. 1984. Biochemistry of bee venom. In: Handbook of natural toxins, Vol. 2, A.T. Tn, (ed.), Marcel Dekker, New York, 732 pp.: 49-85.

Simal, J., Huidobro, J.F. and Muniategui, S. 1988. [Study of the sterol fraction of bee-collected pollen.] Grasas y Aceites, 39(6): 327-333

Simu'th, J., Trnovsky, J. and Jeloskova', J. 1986. Inhibition of bacterial DNA-dependent RNA polymerases and restriction endonuclease by UV-absorbing components from propolis. Pharmazie, 41(2): 131-132

Sista, M., Das, K., Mandal, K. and Gupta, K. 1986. Preparation and in vitro dissolution study of wax-embedded prolonged release dosage form containing chlorpheniramine maleate. Indian Drugs, 24(1): 34 36

Smith, A.K. and Circle, S.J. 1972. Chemical composition of the seed. In: Soybeans:

Chemistry and Technology. Smith and Circle (eds.), Avi Publ. Co., Westport, CT, USA : 61-90

Smith, F.G. 1951. Preliminary report on Trigona wax. E. Afr. Agric. J., 16 (4): 185-187 Smith, F.G. 1954. Notes on the biology and waxes of four species of African Trigona bees (Hymenoptera: Apidae). Proc. R. Ent. Soc. Lond. Ser. A, 29 (4/6):62-70

Spanish Dairy Corporation 1975. Preparation of yoghurt with honey. In: International Symposium on Apitherapy, Madrid, 1974. Apimondia Publishing House, Bucharest, Rumania, p.55-56

Speedy Bee 1988. Miller Honey (Co.) develops dehydrated, pure honey. Speedy Bee, 12 (5): 11

Soldati, A.G. and Piazza, S.M. 1985. Elaboraci6n de hidromiel. Fac. de Agronomia, Univ. de Ia Plata, La Plata, Argentina, unpublished communication.

Sosnowski, Z.M. 1984. Method for extracting propolis and water soluble dry propolis powder obtained thereby and cosmetic and pharmaceutical preparations containing same. European Patent Application, No.0 109 993, 25 pp.

Spitznagel, H.U. 1988. Processing your propolis the right way. Newsl. Can. Hon. Bee Res. Assoc., (Aug.): 5-6

Spöttel, W. 1950. (Honey and dried milk). J.A. Barth, Leipzig, Germany, 323 pp.

Stanley, R.G. and Linskens, H.F. 1974. Pollen: biology, biochemistry, management. Springer Verlag, Berlin, Germany, 307

Stein, I. 1989. Royal jelly: the new guide to nature's richest health food. Thorsons Publishers, Ltd., Wellingborough, U.K., 128 pp.

Steinkraus, K.H. and Morse, R.A. 1966. Factors influencing the fermentation of the honey in mead production. J. Apic. Res. 5:17-26

Subrahmanyam, V.M. 1993. Storage of skin grafts in honey. Bee Well 3(1): 6 (also in Lancet, 1/93)

Szente, L. and Szejtli, J. 1987. Formulation of propolis with B-cyclodextsrin. Acta Pharmaceutica Technologica 33(4): 218-221

Taber, 5. 1991. Bee Management, Part I - Comb Honey Production. Am. Bee J. 131(3):179-180

Tabio, C., Alvarez, J.D. and Berisiartu, M. 1988. [Preliminary characterization of multifloral pollen from the El Cano area of Havana City Province, Cuba.] Ciencia y Te'cnica en la Agricultura, Apicultura, 4: 73-81

Takahashi, M., Matsuo, I. and Ohkido, M. 1983. Contact dermatitis due to honeybee royal jelly. Contact Dermatitis. 9 (6): 452-455.

Takenaka, T. Nitrogen components and carboxylic acids of royal jelly. In Chemistry and biology of social insects (edited by Eder, J., Rembold, H.). Munich, German Federal Republic, Verlag J. Papemy (1987): 162-163.

Takenaka, T., Yatsunami, K. and Echigo, T. 1986. Changes in quality of royal jelly during storage. Nippon Shokuhin Kogyo Gakkaishi 33 (1): 1-7

Tamua, T., Fuj ii, A. and Kubiyama, N. Study on mutagenicity of royal jelly. Honeybee Science (1985) 6 (1): 7-12.

Tamura, T., Fujii, A. and Kubiyama N. 1987. [Antitumor effects of royal jelly.] Nippon, Yakurigaku-Zasshi, 89 (2): 73-80

Taylor, R.L. and Carter, B.J. 1976. Entertaining with Insects, or: The Original Guide to Insect Cookery. Woodbridge Press, Santa Barbara, California, USA, 160 pp.

Telatin, L. Proprieta' terapeutiche della gelatina reale in neuro-psichiatria. Atti del 10 convegno nazionale per lo studio dell'applicazione dei prodotti delle api nel campo medio-biologico, Bologna, Italia (156): 211-214.

Thikonov, A.I. and 3 others 1991. Composition and manufacture of sublingual tablets containing hydrophobic phenolic propolis. Acta Med. Scand., 40 (1): 23-27

Thoenes, S.C. and Schmidt, J.O. 1990. A rapid, effective method of non-destructively removing honeybee larvae from combs. Amer. Bee J., 130: 817

Tikhonov, A.I., Mamontova, IN.S. 1987 [Production and study of a lyophilized phenolic polysaccharide preparation from propolis.] Farmatsevtichnii Zhurnal 3: 67-68

Tikhonov, A.I., Salo, D.P., Pryakhin, O.R. and Gritsenko, V.1. 1978. Standardization of propolis. Pharmaceutical Chem. J., 11(12): 1694-1699

Tong, S.S. C. and 3 others 1975. Elemental analysis of honey as an indicator of pollution: forty-seven elements in honeys produced near highway, industrial and mining areas. Archs Envir. Hith, 30 (7): 329-332

T6th, G., Lemberkovics, E. and Kutasi-Szab6, J. 1987. The volatile components of some Hungarian honeys and their antimicrobial effects. Amer. Bee J., 127 (7): 496-497

Townsend, G.F. and 5 others. 1960. Studies on the in vitro antitumor activity of fatty acids: I. 10-hydroxy-2-decenoic acid from royal jelly. Cancer Res., 20: 5-3-510.

Townsend, G.F. 1975. Processing and storing liquid honey. IN: Honey: a compresensive survey. by Drane, Heinemann, London, U.K.: 269-292

Tulloch, A.P. 1970. The composition of beeswax and other waxes secreted by insects. Lipids, 5 (2): 247-258

Tulloch, A.P. 1973. Factors affecting analytical values of beeswax and detection of adulteration. J. Am. Oil Chem. Soc. 50(7): 269-272

Tulloch, A.P. 1980. Beeswax - composition and analysis. Bee World, 61(2): 47-62

Tumanov, A.A. and Osipova, N.I. 1966. [Biological determination of traces of substances.] Mat. All-Union Conf., 1963, Gorky, USSR, : 238-246

Uccusic, P. 1982. (Doctor Bee: bee products, their curative power and application in medicine.) Doktor Biene: Bienenprodukte - ihre Heilkraft und Anwendung in der Heilkunst. Ariston Verlag, Genf, Switzerland, 2nd edition in 1983, 198 pp.

Unknown 1982. Treatment of timber for hive parts. Apis, South Africa, March : 5-6

USA. 1978. Beeswax. Affirmation of GRAS status as a direct human food ingredient. USA. Laws and Statutes, 14643-14644 Federal Register 43 (68)

Vakikonina, T.V., Duslikova, E.S. and Bodrova, R.N. 1975. Detection of the adulteration of propolis. USSR patent No.474 325

Valdes, G., Rojas, N.M. and Morales, C. 1987. [Preliminary test of the action of propolis extract on Candida albicans.] Ciencia y te'cnica en la Agricultura, Apicultura, 3: 41-49

Vanhaelen, M. and Vanhaelen-Fastre, R. 1979a. [Propolis. - I. Origin, microscopical investigations, chemical constituents and therapeutical activity.] J. Pharm. Belg.. 34 (5): 253-259

Vanhaelen, M. and Vanhaelen-Fastre, R. 1979b. [Propolis. - II. Identification by high performance chromatography (liquid, gas-liquid and thin-layer) of the constituents. Bioautography of the chromatograms of the antibacterial constituents.] J. Pharm. Belg., 34 (6): 317-328

Vecchi, M.A., Sabatini, A.G., Grazia, L., Tini, V. and Zambonelli, C. Il contenuto in vitamine come possibile elemento di caratterizzazione della gelatina reale. Apicoltura (1988) 4:139-146.

Vick, J.A. and Brooks, R.B. 1972. Pharmacological studies of the major fractions of bee venom. Amer. Bee J., 112 (8): 288-289

idyaev, I.P. 1968. (Method of producing preparations for treating wounds and skin diseases in animals). Russ. Patent No.1106348 (1968)

Vinci, L. 1981. The book of practical candle magic. Aquarian Press, Wellingborough, U.K. 128 pp.

Vittek, J. and Halmos, J. 1968. Cytological analysis of the function of the ndosteum and periosteum in reparative osteogenesis of the rabbit mandible following the application of penicillin and royal jelly. Cslka Stomat. 68 (3): 207-216.

Vittek, J. and Slomiany, B.L. 1984. Testosterone in royal jelly. Experientia 40 104-106. Vorwohl, G. 1968. Natiirliche Diastraseschwache der Honige von Apis cerana Fabr. Z. Bienenforsch. 9(5): 232-292

Wakayama, T. and Lee, C.Y. 1987. Factors influencing the clarification of apple juice with honey. Food Chemistry, 25:111-116

Wagner, H., Dobler, I., Thiem, I. Effect of royal jelly on the peirpheral blood and survival rate of mice after irradiation of the entire body with X-rays. Radiobiologia Radiotherapia (1970)11(3): 323-328.

Walker, P. 1983a. Beeswax: uses and commercial aspects. Bibliography, International Bee Research Association, No.33, l7pp.

Walker, P. 1983b. Beeswax: processing. Bibliography, International Bee Research Association, No.31, 11 pp.

Walker, P. and Crane, E. 1987. Constituents of propolis. Apidologie, 18(4): 327-334

Waller, G.D. 1980. A modification of the O.A.C. pollen trap. Amer. Bee J., 120:119-121

Wang, B.J. and Zhang, H.J. 1988. Studies on the chemical constituents of Beijing (China) propolis. Bull. Chinese Materia Medica, 13 (10): 37-38

Wang, W. 1989. The development and utilization of the resources of bee pollen in China. Proc. 32nd Intl. Cong. Apic., Apimondia, (abstract) p.36

Wang, W., Hu, J. and Cheng, J. 1984. (Biological effect of honeybee pollen: radioprotective activity on hemotopoitic tissues of irradiated mice). J. Hangzhou Univ., 11: 231-240

Warth, A.H. 1956. The chemistry and technology of waxes. 2nd ed. Reinhold Publishing Co., New York

Weiss, K. (1975) Zur kastenspezifischen Erna~hrung der Weiblichen Bienenlarve (Apis mellifica L.). Apidologie 6: 95-120.

Weng, C., Hu, Y.C. and Yan, T.C. 1979. A new plant growth regulator (myricyl alcohol) isolated from beeswax. Hua Hsueh Tung Pao, 5: 453-454

Wheatley, et al. 1950. as cited in Klein, 1991

White, E.C. 1993. Super Formulas: arts and crafts. How to make more than 360 useful products that contain honey and beeswax. Valley Hills Press, Starkville, USA. 120 pp.

White, J.W. Jr., Riethof, M.L., Subers, M.H. and Kushnir, I. 1962. Composition of American honeys. Tech. Bull. IU.S. Dept. Agric. 1261, 124 pp.

White, J.W. Jr., Kushnir, I. and Subers, M.H. 1964. Effect of storage and processing temperature on honey quality. Food Technol. 18(4): 154-156

White, J.W. 1966. Improving the colour of beeswax. Gleanings in Bee Culture 94: 742, 743, 758

White, J.W. 1975a. Physical characteristics of honey. In: Honey, a comprehensive survey, Crane (ed.), Heinemann, London, U.K. : 207-239

White, J.W. 1975b. Composition of honey. In: Honey, a comprehensive survey, Crane (ed.), Heinemann, London, U.K. :157-206

White, J.W. 1975c. Honey. In: The hive and the honeybee. Dadant & Sons Inc., Hamilton Illinois, USA, 491-530

White, J.W. and Doner, L.W. 1978. The 13C/12C ratio in honey. J. Apic. Res., 17 ~ 94-99

White, J.W. Jr., Kushnir, I. and Doner, L.W. 1979. Charcoal column/thin layer chromatographic method for high fructose corn syrup in honey and spectrophotometric method for hydroxymethylfurfural in honey: collaborative study. J. Assn. Off. Anal. Chem. 62(4): 921-927

White, J.W. and Siciliano, J. 1980. Hydroxymethylfurfural and honey adulteration. J. Ass. Off. Analyt. Chem., 63 (1): 7-10

White, J.W. 1980. Hydroxymethylfurfural content of honey as an indicator of its adulteration with invert sugars. Bee World, 61(1): 29-37

White, J.W., Platt, Jr. J.L., Allen-Wardell, G. and Allen-Wardell, C. 1988. Quality control for honey enterprises in less-developed areas: an Indonesian example. Bee World, 69 (2): 49-62

Whiterell, P.C. 1975. Other products of the hive. In: The hive and the honeybee, Dadant & Sons, IHamilton, Illinois, USA, : 531-558

Wollenweber, E. and 4 others 1987. A novel caffeic acid derivative and other constituents of Populus bud excretion and propolis (bee-glue). Z. f. Naturf., 42:1030-1034

Yanishlieva, N. and Marinova, E. 1986. [Application of a new method registering propolis components with anti-oxidative effects.] Khranitelnopromishlena Nauka, 2 (3): 45-50

Yanishlieva, N., Marinova, E. and Antonova, V. 1986. Possibilities of enhancing sunflower oil oxidation stability by adding natural antioxidants (Propolis). (In: Chem. Abs. 1986, 105, 17 # 151751h). Khranitelnoprom Nauka, 2 (2): 15

Yatsunami, K. and Echigo, T. 1985. Antibacterial activity of royal jelly. Bulletin of the Faculty of Agriculture, Tamagawa University No.25, 13-22.

Yener, E., Ungan, S. and Ozilgen, M. 1987. Drying behaviour of honey starch mixtures. J.Food. Sci., 52 (4): 1054-1058

Ziai, M.R. and Blume, A.J.H. 1990. Mast cell degranulating peptide: a multi-functional neurotoxin. J. Pharm. Pharmacol. 42(7): 457-461

Zommer-Urbanska, S., Gniazdowski, R. and Bojarowicz, H. 1989. Working out the technology of propolis unguentum and its application in vas-motor catarrh treatment. Proc. 31st Intern. Congr. Apic., Warsaw, Poland, August 1987: 488-492

Zwaenepoel, C. 1984. Honey: facts and folklore. Alberta Beekeepers' Association, Edmonton, Canada, 24 pp.

Index